Emissions Trading

Emissions Trading
an exercise in
reforming pollution policy

T. H. Tietenberg

RESOURCES FOR THE FUTURE, INC. / WASHINGTON, D.C.

Published by Resources for the Future, Inc., 1755 Massachusetts Avenue, N.W.,
Washington, D.C. 20036
Resources for the Future books are distributed worldwide by The Johns Hopkins
University Press.

Library of Congress Cataloging in Publication Data

Tietenberg, Thomas H.
  Emissions trading, an exercise in reforming pollution policy.

  Includes index.
  1. Air—Pollution—Government policy—United States.
I. Title.
HC110.A4T54   1985   363.7′39256′0973   84-18335
ISBN 0-915707-12-8

This book was produced under an RFF Gilbert White Fellowship. It was designed by Elsa B. Williams and edited by Ruth B. Haas. The index was prepared by Lorraine and Mark Anderson.

Contents

Preface   xi

Abbreviations   xiv

1 / Introduction   1

The Policy Context   2
The Emissions Trading Program   7
An Overview of the Book   11

2 / The Conceptual Framework   14

The Regulatory Dilemma   14
Cost-Effective Permit Markets   16
Other Properties of Transferable Permit Systems   30
Summary   35

3 / The Potential for Cost Savings   38

The Nature of the Evidence   39
The Magnitude of Potential Cost Savings   41
The Potential for Actual Cost Savings   47
Summary   56

vii

4 / The Spatial Dimension 60

Difficulties in Implementing an Ambient Permit System 60
Possible Alternatives 64
The Current Emissions Trading Program 86
Summary 89

5 / Distributing the Financial Burden 93

The Command-and-Control Financial Burden 94
Assigning the Initial Control Responsibility 97
The Size of the Financial Burden 102
The EPA Emissions Trading Program 113
Summary 120

6 / Market Power 125

Permit Price Manipulation 126
Reducing Competition 138
The EPA Emissions Trading Program 141
Summary 145

7 / The Temporal Dimension 149

The Probabilistic Nature of Pollutant Concentrations 149
Cost-Effective Temporal Control 152
The EPA Emissions Trading Program 160
Summary 165

8 / Enforcement 168

The Nature of the Enforcement Process 169
The Economics of Enforcement 171
Current Enforcement Practice 175
Summary 184

9 / Evaluation and Proposals for Further Reform 188

Evaluation 188
Proposals for Further Reform 202
Proposals for Future Research 212
Concluding Comments 213

Index 217

Tables

1. National Primary and Secondary Ambient Air Quality Standards *4*
2. The Evaluation of the Emissions Trading Program: Major Milestones, 1975–84 *10*
3. Cost Effectiveness for Nonuniformly Mixed Assimilative Pollutants: A Hypothetical Example *25*
4. Empirical Studies of Air Pollution Control *42*
5. Empirical Studies of Water Pollution Control *46*
6. Bubble Trades Approved by EPA During 1981 *54*
7. Bubble Transactions by Type of Pollutant *56*
8. A Comparison of Emission Reduction for Command-and-Control and Least-Cost Approaches *65*
9. Using Emission Permit Systems to Control Nonuniformly Mixed Assimilative Pollutants: The Potential Cost *68*
10. Correlation of Ranks Between Relative Level of Emission Reduction and Control Cost: Command-and-Control and Emission Permit Systems *71*
11. The Effect of the Number of Zones on the Potential Cost Effectiveness of a Full-Information Zonal Permit Policy: Particulate Control in Baltimore, Maryland *77*
12. The Size of the Potential Regional Financial Burden: Air Pollutants *104*
13. The Size of the Potential Regional Financial Burden: Water Pollution *107*
14. Competitive and Noncompetitive Auction Markets: A Numerical Example *129*
15. Noncompetitive Function Markets and Financial Burden: The Mohawk River *132*

Preface

As anyone who has tried it knows, regulatory reform is easier said than done. Reform concepts which appear so disarmingly simple in the abstract world of theory turn out to be distressingly complex when applied. Regulations which from a distance seem so inherently unsupportable, upon closer inspection are discovered to have significant bases of support among various special interest groups. Since the status quo has so much inertia, many promising ideas end up strewn along the wayside. Survivors are few and far between.

What is the price of survivorship? How much of the original idea has to be sacrificed as the cost of gaining a place in the sun? One way to begin to answer these questions is to examine closely those reform packages, such as the emissions trading program, that have survived.

From my perspective, the emissions trading program is a particularly interesting example of a survivor because it alters the way regulators control air pollution. Historically, the political process had not only been insensitive to the point of view that environmental policy could be improved by a greater reliance on economic incentives, but, on those few occasions when serious attempts were made to incorporate economic incentives, it was downright hostile. In light of this hostility, the enduring role forged for the emissions trading program, an approach that relies heavily on economic incentives, is all the more remarkable.

In this book I attempt to evaluate the emissions trading program using two different benchmarks: (1) Did its introduction improve upon the policy that preceded it? and (2) How closely did it fulfill the basic objectives which motivated the program? The former benchmark is particularly helpful in establishing whether this reform represented progress, while the latter benchmark helps to identify and, to the extent possible, quantify deviations from the original policy goals caused either by unanticipated implementation complexities or by intentional sacrifices made to build political support for the idea. Not so coincidentally, the latter benchmark also provides a means for creating a menu of possible opportunities for moving the program even closer to fulfilling its original promise.

In the thirteen years I have been learning and writing about emissions trading programs in general and the EPA emissions trading program in particular, I have discovered that interest in this subject comes from a number of different quarters. Economists see this program not only as a rationalization of existing policy, but also as a harbinger of things to come in other areas of environmental policy and in other parts of the world. State, regional, and national regulators who have invested a considerable amount of time in grappling with the day-to-day operations of the program have expressed an interest in exploring how the pieces have (and should have) fit together as well as in ascertaining how typical their own experiences with the program have been. Through their writings and interviews, lawyers have articulated a particular interest in the legal evolution of the program as determined by the bureaucracy operating within the confining influences of Congress and the courts. Though diverse, these are related, even complementary, interests.

In recognition of the rather diverse backgrounds of potential readers, I have tried to write a book which, with the possible exception of portions of chapter 2, is accessible to a general, but informed audience. Even chapter 2, which contains some potentially intimidating mathematics, is written so the nonpurist can gain the insights conveyed by the math while skipping the equations. Because I have learned so much from noneconomists as well as from economists, I have attempted to write a book which is rooted in economics, but which ultimately transcends disciplinary boundaries in both style and substance.

In writing this book I have been aided by a large number of people. My greatest debt of gratitude is owed to Clifford S. Russell and Allen V. Kneese, both of Resources for the Future, who read the entire manuscript and gave me the benefit of their very helpful comments. Others who read and made valuable comments on portions of the manuscript include Walter O. Spofford, Jr., Winston Harrington, and Paul Portney at Resources for the Future; Robert W. Hahn (Carnegie–Mellon University); Randolph M. Lyon (University of Texas–Austin); Albert M. McGartland and Mike Levine of the U.S. Environmental Protection Agency; Scott E. Atkinson (University of Wyoming); and William B. O'Neil (Colby College).

Others contributed by allowing me to pick their brains during extended conversations. These included David Doniger (Natural Resources Defense Council); Leslie Sue Ritts (Environmental Law Institute); John Palmisano (private consultant); and Ivan Tether, David Foster, and Indur Goklany, all with the U.S. Environmental Protection Agency.

Still others were kind enough to provide me with unpublished material or material not readily available in the Washington area. These included Alan Krupnick (Resources for the Future), Jay Narco and Kevin Croke (ETA Engineering), William J. Baumol (New York University and Princeton University), David Harrison (Harvard University), Lloyd Kostow (Oregon Department of Environmental Protection), Henry Droege (Washington Department of Ecology), Chris Romaine (Illinois Environmental Protection Office), Robert Yeates (New Jersey Department of Environmental Protection), Thomas Getz (Rhode Island Department of Environmental Management), Fred Baumberger (California Air Resources Board), David Dixon (Maine Department of Environmental Protection), and Elliot Gilberg (U.S. Environmental Protection Agency).

A sabbatical research grant from Colby College and a Gilbert White Fellowship from Resources for the Future made completion of this book possible.

Portions of this manuscript were expertly typed by Betty Cawthorne, D. J. Curran, Anne Farr, Pat Flynn, and Margaret Parr-Recard. Marilyn Voigt coordinated and managed the process of turning my handwritten copy into a polished manuscript.

I am deeply grateful for all these contributions. I wouldn't have been able to do it without them.

Washington D.C.   Tom Tietenberg
July 1984

ABBREVIATIONS

SIP	State implementation plan
LAER	Lowest achievable emission rate
PSD	Prevention of significant deterioration
BACT	Best available control technology
NSPS	New source performance standards
RACT	Reasonably available control technology
TDP	Transferable discharge permit

1 / Introduction

Although there has been a long, active, intellectual tradition associated with regulatory reform, only in the past ten to fifteen years has popular support grown sufficiently to place it high on the political agenda. During the beginning of the 1970s, President Nixon established the Advisory Council on Executive Organization to study the independent regulatory agencies and to recommend organizational improvements.[1] President Carter carried this initiative a step further by establishing both the Regulatory Analysis Review Group and the Regulatory Council within the executive branch to identify good reform ideas and to facilitate their implementation.[2] Apparently the voters felt even these measures were not enough as President Reagan was propelled into the White House in 1980, in part because of his promise to further lessen the burden of regulation.[3]

Despite this apparently intense interest, there are remarkably few reform proposals which have successfully negotiated the perilous path from concept to implementation. Certainly this is not due to any shortage of ideas on how existing regulations or the regulatory process could be improved. The literature is full of proposals.[4] Even if only a fraction

1. For a description and evaluation of the work of this council, see Noll (1971).
2. The roles of these organizations and the issues they addressed are described in White (1981).
3. Regulatory reform as an issue in the Carter–Reagan campaign is explored in Stone (1982, pp. 238–239).
4. For a sample of this literature, see Stephen Breyer, *Regulation and Its Reform* (Cambridge, Mass., Harvard University Press, 1982); Leroy Graymer and Frederick

1

of the proposals were to prove meritorious, the number of implemented reforms would be insignificant compared with the number of ideas.

The paucity of implemented reforms suggests that there may be much to learn from those that did become policy. One leading example is the emissions trading program. Hailed by Senator Pete Domenici (R. New Mexico) as "the one bright idea that has emerged in the 1980's,"[5] this program was created by the U.S. Environmental Protection Agency (EPA) to inject more flexibility into the regulatory procedures for meeting air quality goals set by the Clean Air Act. To understand the role played by this reform in environmental policy, it is necessary to have some familiarity with the manner in which the Clean Air Act regulates stationary sources.

THE POLICY CONTEXT

Pollutant Types

The Clean Air Act currently recognizes two main types of air pollutants, *criteria* pollutants and *hazardous* pollutants. Each type has a unique control policy. The seven criteria pollutants[6] are relatively common substances, are found in almost all parts of the country, and are presumed to be dangerous only in high concentrations. Hydrocarbons (also referred to as volatile organic compounds) are included primarily because they contribute to the creation of ozone, one of the other pollutants. Lead, the most recent addition, was included as the result of a lawsuit brought by the Natural Resources Defense Council in 1975.

These pollutants are called "criteria pollutants" because the act requires EPA to produce "criteria documents" to be used in setting acceptable levels for these pollutants. They summarize and evaluate all of the existing research on the health and environmental effects associated with these pollutants. Historically, the central focus of the Clean Air Act has been on criteria pollutants, though the emphasis on hazardous pollutants is growing.

Thompson, eds., *Reforming Social Regulation: Alternative Public Policy Strategies* (Beverly Hills, Calif., Sage Publications, 1982); Robert E. Litan and William D. Nordhaus, *Reforming Federal Regulation* (New Haven, Conn., Yale University Press, 1983); and Wesley A. Magat, ed., *Reform of Environmental Regulation* (Cambridge, Mass., Ballinger, 1982).

5. *Congressional Record-Senate* (December 15, 1982): S14668.

6. The criteria pollutants are sulfur dioxide, total suspended particulates, carbon monoxide, nitrogen oxides, hydrocarbons (also referred to as volatile organic compounds), ozone, and lead.

The second class of pollutants contains a number of airborne substances which have been implicated in cancer, genetic damage, neurotoxicity, reproductive, or other serious health effects. Unlike the criteria pollutants, small doses of hazardous pollutants can produce serious damage. The Clean Air Act empowered EPA to list and regulate any pollutants that fit this description. As of 1983, only seven had been listed.[7] Whereas criteria pollutants exist virtually everywhere and affect large numbers of people, hazardous pollutants are found only in certain locations, with substantially lower numbers of the population exposed. Because they are toxic, however, they have the potential of being more harmful to those exposed than similar doses of the criteria pollutants.

Air Quality Standards

For each criteria pollutant, the Environmental Protection Agency has set legal ceilings (called ambient air quality standards) on concentrations of the pollutant in outdoor air over some time period. Many pollutants have the standard defined in terms of either a long-term average (defined normally as an annual average) or a short-term average (e.g., a 3-hour average), or both. The ambient standards as of 1982 are given in table 1. By law, they have to be reviewed every five years and modified as necessary. These standards have to be met everywhere, though as a practical matter they are monitored at a large number of specific locations.

The *primary standard* is designed to protect human health and has the earliest deadlines for compliance. All pollutants have a separate primary standard. The *secondary standard* is designed to protect aspects of human welfare other than health where separate effects have been observed for specific pollutants. These include aesthetics (particularly visibility) and damage to physical objects (e.g., houses, monuments, etc.), and vegetation. As of 1983, only sulfur dioxides and total suspended particulates had separate secondary standards. Because the secondary standard is normally more stringent than the primary, once the deadline for compliance with the secondary standard has been reached, it, rather than the primary standard, tends to govern the degree of control required.

State Implementation Plans

While EPA has the responsibility for defining ambient standards, the primary responsibility for ensuring that the ambient air quality standards

---

7. The listed hazardous pollutants are asbestos, beryllium, mercury, vinyl chloride, benzene, radionuclides, and arsenic.

Table 1.   National Primary and Secondary Ambient Air Quality Standards

Pollutant	Primary standard	Secondary standard
Sulfur oxides	a. 80 μg/m^3 (0.03 ppm) annual arithmetic mean b. 365 μg/m^3 (0.14 ppm) maximum 24-hour concentration not to be exceeded more than once a year	1,300 μg/m^3 maximum 3-hour concentration
Particulate matter	a. 75 μg/m^3—annual geometric mean b. 260 μg/m^3—maximum 24-hour concentration not to be exceeded more than once per year	a. 60 μg/m^3 annual geometric mean b. 150 μg/m^3—maximum 24-hour concentration not to be exceeded more than once per year
Carbon monoxide	a. 10 mg/m^3 (9 ppm) maximum 8-hour concentration not to be exceeded more than once per year b. 40 mg/m^3 (35 ppm) maximum 1-hour concentration not to be exceeded more than once per year	No separate secondary standard
Ozone	235 μg/m^3 (0.12 ppm) maximum average hourly concentration not to be exceeded more than once a year	No separate secondary standard
Hydrocarbons	160 μg/m^3 (0.24 ppm) maximum 3-hour concentration (6 to 9 a.m.) not to be exceeded more than once a year	No separate secondary standard
Nitrogen dioxide	100 μg/m^3 (0.05 ppm) annual arithmetic mean	No separate secondary standard
Lead	1.5 μg/m^3 arithmetic mean averaged over a calendar quarter	No separate secondary standard

Notation: μg/m^3 is micrograms per cubic meter
mg/m^3 is milligrams per cubic meter
ppm is parts per million
   Source: Code of Federal Regulations vol. 40, parts 50.4, 50.6, 50.7, 50.8, 50.9, 50.10, 50.11 and 50.12 (1982).

are met falls on the state air pollution control agencies. They develop and execute an acceptable state implementation plan (SIP), which must be approved by the EPA. Once approved, it can be enforced by EPA as well as by the states. The SIP spells out, for each separate air quality control region, procedures and timetables for meeting ambient standards. The degree of control required by these plans depends on the severity of the pollution problem in each of the control regions.

NONATTAINMENT REGIONS. By 1975 it had become apparent that despite some major gains in air quality, many areas had not met, and would not meet, the ambient standards for certain pollutants by the statutory deadlines. Therefore in the 1977 Amendments to the Clean Air Act, Congress extended the deadline for attainment of all primary (health-related) ambient standards to 1982, with further extensions to 1987 possible for ozone and carbon monoxide. The amendments also required EPA to designate all areas not currently meeting the standards as *nonattainment regions*. These areas were subjected to particularly stringent controls. After the 1977 Amendments were passed, all portions of state implementation plans applying to nonattainment regions had to be revised by the state control authorities to demonstrate compliance with the new deadlines. Congress gave EPA the power to halt the construction of major new projects or major expansions of existing plants and to deny federal sewage and transportation grants to any state not submitting a plan demonstrating how attainment would be reached by the statutory deadlines.

The statutes specifically call for these plans to provide for implementation of all "reasonably available control measures as expeditiously as practicable" on all existing sources of air pollution and for "reasonable further progress" on an annual basis toward meeting the standards. The former requirement mandates stringent emission standards (called RACT standards) for existing sources while the latter requires annual reductions in emissions of a size that guarantees compliance by the deadline.

State implementation plans in nonattainment regions must also include a construction permit program for major new sources or large sources undergoing some major modification. Permits cannot be granted to these sources unless the state can demonstrate that the emissions resulting from the proposed changes would not jeopardize the region's progress toward attainment. The state can satisfy this requirement by ensuring that controls on existing sources are stringent enough to allow progress to be demonstrated even with the new sources in operation.

A second condition for the permit stipulates that all major new or modified sources in nonattainment areas must also control their own emissions to the "lowest achievable emission rate." The *lowest achievable emission rate* (LAER) is defined as the lowest emission rate achieved by any similar source anywhere in the country, or the lowest emission rate for a similar source included in any state implementation plan. This mandatory minimum control level for all new sources was designed to ensure that only the most stringent controls would be used by any source locating in a nonattainment area.

PSD REGIONS. Regions with air quality better than the ambient standards were subject to another set of controls known collectively as the PSD policy. This policy derives its name from its objective, namely the prevention of *significant deterioration* of the air in cleaner regions. The origins of this policy can be traced back to a section in the preamble to the 1970 Clean Air Act, which stated as an objective: "to protect and enhance the quality of the nation's air." In 1972 the EPA was successfully sued by the Sierra Club for promulgating regulations which failed to ensure the achievement of this objective. The court held that while the ambient standards prevented the deterioration of the air *beyond* the standard, nothing in the regulations prevented the air significantly cleaner than the standard from deteriorating until it reached the standard. Following the court's decision, EPA adopted a PSD program in 1974 and the 1977 Amendments to the Clean Air Act continued a modified version of that program.

The PSD regulations specify the maximum allowable increases or increments in pollution concentration beyond a historically defined baseline. New sources seeking to locate in PSD regions must secure permits. As one condition of securing their permits, these sources must install the *best available control technology* (BACT). Just what technologies satisfy this requirement are determined by states on a case-by-case basis. Each new source built consumes a portion of the allowable increment. Once the increment has been completely consumed, no further deterioration of the air is allowed in that area, even if the air were cleaner than required by the prevailing ambient standard. For all practical purposes, whenever the PSD increments are binding they define a tertiary standard which varies in magnitude from region to region.

National Emission Standards

In addition to defining the ambient standards and requiring states to define BACT and LAER emission standards, the EPA has itself established national, uniform emission standards for two categories of stationary sources: (1) those emitting hazardous pollutants, and (2) new sources of criteria pollutants or major modifications of existing sources. The hazardous pollutant standards are preemptory; the EPA has assumed direct responsibility rather than delegating it to the states. The standards governing new and modified stationary sources of criteria pollutants, the *new source performance standards* (NSPS), were designed to serve as a floor for BACT and LAER determinations by the states. Congress wanted to establish a uniform national floor for the amount of required emission control from new sources in order to prevent states from caving in under industry pressure. Neither LAER nor BACT can be lower than the new source performance standards.

THE EMISSIONS TRADING PROGRAM

Stripped to its bare essentials, the Clean Air Act's approach toward stationary sources involves the specification of emission standards (legal ceilings) on all major emission sources. These standards are imposed on a large number of specific emission points such as stacks, vents, or storage tanks.

The emissions trading program attempts to inject more flexibility into the manner in which the objectives of the Clean Air Act are met. Sources are encouraged to change the mix of control technologies envisioned in the standards as long as air quality is improved or at least not adversely affected by the change. The program is implemented by means of four separate policies, linked by a common element known as the emission reduction credit. The emission reduction credit is the currency used in trading among emission points, while the offset, bubble, emissions banking, and netting policies govern how the currency can be spent.

The Components of the Program

THE EMISSION REDUCTION CREDIT. Should any source decide to control any emission point to a higher degree than necessary to fulfill its legal obligations, it can apply to the control authority for certification of the excess control as an emission reduction credit. Certified credits can be banked or used in the bubble, offset, or netting programs. To receive certification, the emission reduction must be: (1) surplus, (2) enforceable, (3) permanent, and (4) quantifiable.

THE OFFSET POLICY. The offset policy was established to resolve a conflict between economic growth and progress toward meeting the ambient standards in nonattainment areas. The dilemma posed by this conflict involved how new or expanded sources could be accommodated while meeting the statutory requirement that the ambient standards be met as expeditiously as possible. Since these sources would add emissions to the region, some means of offsetting them had to be found.

The offset policy allows qualified new or expanding sources to commence operations in a nonattainment area provided they acquire sufficient emission reduction credits from existing sources. By buying the credits, new sources, in effect, finance emission controls undertaken by existing sources. This approach was designed to ensure that regional emissions would be lower after the source began operations (counting the acquired emission reduction credits) than before. Major new or modified sources are qualified to participate in this program only if they control their own emissions to the degree required by the LAER standard and all existing major sources owned or operated by the applicant in the same state as the proposed source are in compliance with their legal control responsibilities.

THE BUBBLE POLICY. The bubble policy allows existing sources to use emission reduction credits to satisfy their SIP control responsibilities. For example, existing sources in nonattainment areas can meet their assigned RACT standards either by adopting the control technology used to define the standard or by adopting some technology that emits the pollutant at a somewhat higher rate, making up the difference with acquired emission reduction credits. The sum of emission reduction credits plus actual reductions must equal the assigned reduction.

This policy derives its unusual name from its treatment of multiple emission points as if they were contained within an imaginary bubble, regulating only the amount leaving the bubble. These bubbles can be extended to include not only emission points within the same plant, but emission points in plants owned by other firms as well.

NETTING. Netting allows sources undergoing modification or expansion to escape the burden of new source review requirements so long as any net increase (counting the emission reduction credits) in plantwide emissions is insignificant. Traditionally, the test of whether a source was subject to the new source review process or not was applied by calculating the expected increases in emission occurring after modernization or expansion. When these increases passed predetermined thresholds, the source was subject to review. Netting allows emission reduction credits earned elsewhere in the plant to offset the increases expected from the expanded or modernized portion in order to determine whether the threshold had been exceeded. By "netting out" of review, the facility may be exempted from the need to acquire preconstruction permits as well as from meeting the associated requirements, such as modeling or monitoring the impact of the new source on air quality, installing BACT or LAER control technology, or meeting the offset requirement; it may also avoid any applicable bans on new construction. Those facilities satisfying the significant increase threshold must still meet emission limits established by the NSPS. Emission reduction credits cannot be used to avoid this national standard.

BANKING. The banking component of the emissions trading program establishes procedures that allow firms to store emission reduction credits for subsequent use in the bubble, offset, or netting programs. States are authorized to design their own banking programs as long as the rules specify the ownership rights over the banked credits; the sources eligible to bank emission reduction credits; and the conditions governing the certification, holding, and use of these credits. The rules must be consistent with the need to meet the requirement in the Clean Air Act for reasonable further progress.

The Evolution of the Program

Since its inception, the emissions trading program has been in constant flux (table 2). The program got off to an inauspicious start. The original version of the bubble program, promulgated in 1975, was designed to allow modified sources greater flexibility in how they were to meet the NSPS requirements. The courts ruled that EPA had exceeded its authority and voided the regulations.

The offset policy came next. Faced with the unpleasant prospect of prohibiting new sources from entering control regions with air quality worse than the standards, EPA used its regulatory authority to establish the offset policy as a way of allowing growth while assuring emission reductions for the region as a whole. When it was writing the 1977 Amendments to the Clean Air Act, Congress wrote most of the PSD program, including the offset program, into the statute. Following these amendments, the offset policy was the only component of the program to be specifically authorized by statute. Though based on general principles in the statutes, the rest remain purely administrative creations.

Prior to the 1977 Amendments, the offset policy ruled out emissions banking on the grounds that banking credits for subsequent use would introduce a conflict with the need to achieve the ambient standards as quickly as possible. Following the 1977 Amendments, EPA reversed its position and allowed banking, concluding that the conflict had been resolved once Congress established specific procedures for reaching attainment which were compatible with emissions banking.

A reincarnated bubble policy with a quite different focus was then initiated as a means of allowing *existing* sources some flexibility in meeting their responsibilities. Prior to this version of the bubble policy, only major new or modified sources could use emission reduction credits.

Netting was the last component to be added. It was born out of a widespead feeling that the new source review process was excessively cumbersome, particularly for modified sources.[8]

If it had been fully implemented when promulgated, this program could have exempted a large number of modified sources from review,

8. See the findings of the National Commission on Air Quality (1981) on the complexity of the permitting process.

Table 2. The Evolution of the Emissions Trading Program: Major Milestones, 1975–84

Date and citation	Title	Significance
December 16, 1975 40 FR 58416	Standards of performance for new stationary sources	First use of bubble concept. Would excuse modified plants from NSPS so long as total emissions do not increase.
December 21, 1976 41 FR 55254	Emissions offset interpretive ruling	Initiated offset policy.
August 7, 1977 91 Stat. 712	Amendments to Clean Air Act	Statutory recognition of offset policy.
January 27, 1978 578 F.2d 319	*ASARCO Inc.* v. *EPA*	Struck down 1975 NSPS bubble.
January 16, 1979 44 FR 3274	Emission offset interpretive ruling (revised)	Revised offset policy to conform with 1977 Amendments. Allowed banking.
January 18, 1979 44 FR 3740	Recommendation for alternative emission reduction options within state implementation plans	Proposed rules establishing bubble policy.
December 11, 1979 44 FR 71780	Recommendations for alternative emission reduction options within state implementation plans	Final bubble rules.
August 7, 1980 45 FR 52676	Requirements for preparation, adoption, and submittal of implementation plans and approval and promulgation of implementation plans	Separate netting rules established for PSD and nonattainment areas.
April 6, 1981 46 FR 20551	Approval and promulgation of state implementation plans: New Jersey	Approved New Jersey's generic VOC bubble; encouraged other states to develop generic rules.
October 14, 1981 46 FR 50766	Requirements for preparation, adoption, and submittal of implementation plans and approval and promulgation of implementation plans	Netting rules changed to provide for uniform treatment of sources in attainment and nonattainment areas.
April 7, 1982 47 FR 15076	Emissions trading policy statement: general principle for creation, banking, and use of emission reduction credits	Integrated bubble, offset, banking, and netting into a single emission trading program.
June 25, 1984 52 LW 4845	*Chevron U.S.A.* v. *Natural Resources Defense Council, Inc.*	Upheld use of netting rules for nonattainment as well as PSD regions.

but it was challenged in the courts by the Natural Resources Defense Council. Stating that exempting modified sources from review in non-attainment areas was inconsistent with the statutory intent to reach attainment as expeditiously as possible, the appeals court voided the netting rules as they apply to sources in nonattainment areas. Ultimately this decision was overruled by the U.S. Supreme Court[9] and the netting program was given the green light in nonattainment as well as PSD regions.

Another recent change in the program involves the use of generic state rules in the bubble policy. Since the emissions trading program is voluntary, states will choose to use it only if they perceive that the benefits outweigh the costs. Originally the bubble policy could be used only if the approving state included the intended trade in a formal revision of its SIP. Because the SIP approval process is the primary means by which EPA exercises its responsibility for assuring state compliance with the Clean Air Act, SIP revisions are bureaucratically cumbersome. Because any SIP revision has to fulfill a large number of procedural requirements, state control authorities are reluctant to file revisions except when absolutely necessary. Requiring bubble trades to be approved through SIP revisions was a surefire way to kill any interest state control authorities might have had in the program.

During 1980, EPA significantly lowered this procedural burden by approving a generic volatile organic compound bubble rule drawn up by New Jersey. Other states were invited to follow suit and several have. Though initially this invitation included only volatile organic compounds, it was subsequently extended to include other criteria pollutants as well. By approving in advance the generic rules states intended to use to govern possible trades, EPA eliminated the need for states to obtain approval of SIP revisions for each bubble trade, simplifying the approval process greatly.

AN OVERVIEW OF THE BOOK

This reform package did not command an immediate constituency and building one was not easy. Even industrial sources, the most natural constituents in light of their potential cost savings and the flexibility of the program, were far from unanimous in their enthusiasm. To some extent they feared that this flexibility entailed greater risk. When an EPA-recommended technology failed to live up to standards, the source

9. The lower court ruling was issued in *Natural Resources Defense Council, Inc. v. Gorsuch*, 685 F 2d 718 (1982). The opinion in the Supreme Court case is summarized in *Chevron U.S.A., Inc. v. Natural Resources Defense Council, Inc.*, 52 LW 4845

could claim it had lived up to its responsibilities, but when the control mix was up to the source, it would lose this defense. Similarly, by reducing emissions more than required by law in order to gain emission reduction credits, plants could alert control authorities to the fact that additional control was possible. Should control authorities use this information to revise the control baseline upward, similar plants under the same ownership would be adversely affected by the creation of the credits. Industrialists could end up losing more in the long run than they gained in the short run.

Though this was a voluntary program, requiring the full cooperation of state control agencies, state cooperation was not inevitable. State authorities feared that the new programs would be more difficult to administer and saw them as a threatening departure from their comfortable, customary way of doing business. Environmentalists feared, and no doubt some industrialists hoped, the program would open a large number of loopholes, leaving a legacy of reduced compliance.

Despite this opposition, the program exists and is likely to continue to exist for some time. Since so many ideas for reform failed to come to fruition, it is natural to ask what sacrifices were made to place this program on the books. Compromise is, after all, the essence of most successful reforms. To what extent were the stated goals of the emissions trading program—increased cost effectiveness and increased speed of compliance—compromised as the price of initiating and maintaining the program?

Answering this question presupposes a benchmark against which the program can be measured. Two rather different benchmarks suggest themselves. The first, the situation prior to the implementation of the program, is particularly helpful in establishing the extent to which the program represented an improvement in air pollution control policy. The second, a theoretically optimal program which is fully cost effective and yields the maximum speed of compliance, serves to define a menu of possible further reforms. Both are useful standards in their own ways.

The evidence on which this evaluation is based is drawn from three rather different sources. Economic theory, the first source, is used to define cost effectiveness, to derive the conditions any cost-effective allocation must satisfy, to show how these conditions depend on the nature of the pollutant being controlled, and to identify the attributes any cost-effective emissions trading program must exhibit. Computer simulations, the second source, flesh out the bare bones of theory. They permit an investigation of how large the control cost differences between the command-and-control and emissions trading allocations are as well as identification of the determinants of the magnitudes of those differences. By incorporating the specific meteorological and source configuration characteristics unique to each region studied, computer simulations bring the general results of theory into sharper focus. The final source, actual emissions trading transactions, allows us to study in some detail how the program has worked in practice. Areas singled out for special attention include the degree to which costs have been reduced, compliance with the regulations has been improved, pollution concentrations have been lowered, and technological progress has been stimulated.

The remainder of the book is divided into three parts. The first part, consisting of chapters 2 and 3, lays the groundwork for the detailed analysis in the succeeding chapters by examining the concept of emissions trading unencumbered by implementation details. Chapter 2 develops the theory behind the emissions trading program, while chapter 3 estimates the magnitude of potential cost savings that could accrue to well-designed applications of this concept.

The next portion of the book is concerned with evaluating the manner in which the emissions trading program has coped with a number of practical implementation problems. Chapter 4 opens this section by examining how the location of the emissions is treated in emission trades. Since the financial burden borne by sources turns out to be quite sensitive to the pretrade allocation of control responsibility, chapter 5 explores this issue in some depth. The last three chapters in this section cover the potential for and consequences of market power, emissions timing, and enforcement of the permits.

The final section of the book, chapter 9, weaves together the insights gained from the individual topic-by-topic examinations of the program to form a comprehensive evaluation. This evaluation serves as the basis for a series of specific recommendations on how the program could be improved.

REFERENCES

National Commission on Air Quality. 1981. *To Breathe Clean Air* (Washington, D.C., U.S. Government Printing Office).

Noll, Roger. 1971. *Reforming Regulation: An Evaluation of the Ash Council Proposals* (Washington, D.C., Brookings Institution).

Stone, Alan. 1982. *Regulation and Its Alternatives* (Washington, D.C., Congressional Quarterly Press).

White, Lawrence J. 1981. *Reforming Regulation: Processes and Problems* (Englewood Cliffs, N.J., Prentice-Hall).

2 / The Conceptual Framework

Why have emission reduction credits become the centerpiece of environmental policy reform? To answer this question, as well as to provide a basis for evaluating how successful the reforms have been, some notion of an optimal mechanism for allocating control responsibility must be defined. This mechanism can then serve as one of the benchmarks against which the existing system can be measured.

As is well known in the economics literature, the mechanism that is most closely related to the emissions trading program is the transferable discharge permit (TDP) market. Under fairly general conditions TDP markets encourage rapid compliance with a cost-effective allocation of the control responsibility. Because the emissions trading program created by EPA is a special case of this more general approach, the large amount of analysis which has been directed toward the former can be used profitably to understand, evaluate, and create an agenda for reform of the latter.

THE REGULATORY DILEMMA

There are two principal participants in the process to regulate the amount of pollution in the nation's air. While the regulatory authority has the statutory responsibility for meeting pollution targets, the human sources of the pollutant (such as industries, automobile drivers, etc.) must ultimately take the actions which will reduce pollution sufficiently to meet the target.

The main responsibilities of the regulator are to decide how to allocate control responsibility among the sources, to design the regulations implementing the decision, and to enforce the resulting regulations. This is a challenging responsibility. Since every home furnace is a pollution source, the number of sources is extremely large. Even the number of major stationary sources is large. It has been estimated, for example, that there are 27,000 major sources of air pollution in the United States.[1]

Traditionally, the regulatory authority has gone about its job by establishing separate emission standards for each point of discharge from major sources of the pollutant. For ease of reference, this means of distributing control responsibility among points of discharge will henceforth be called the *command-and-control approach*. Since each industrial plant will typically contain several pollutant discharge points, each with its unique emission standard, the amount of information the control authority would need if it were to define cost-effective standards is staggering. Typically, the amount of information available to it when the allocations are made falls far short of what is needed for this task.

In some ways the managers of plants emitting pollution are in exactly the opposite position. Because each plant manager typically would know the unique array of possible control techniques most suited to his or her operation, as well as the associated costs and reliability of these techniques, the quality of information at this level of decision-making is very good. Plant managers generally will have a very good feel for which control technologies would produce the most cost-effective emission reductions at their plants.

Unfortunately, however, plant managers lack the incentive to act on this information in a manner consistent with cost-effective emission reduction. Since any unilateral increase in cost incurred by individual plants faced with competition either from existing or potential rivals could weaken their competitive position, plants would seek to minimize their own costs using any means at their disposal. Possible means include overstating costs to the control authority or to the legislature in hopes of being allocated a weak standard, or seeking an exemption from the courts on grounds of affordability or technological infeasibility.

The fundamental problem with the command-and-control approach is a mismatch between capabilities and responsibilities. Those with the incentive to allocate the control responsibility cost effectively, the control authorities, have too little information available to them to accomplish this objective. Those with the best information on the cost-effective choices, the plant managers, have no incentive either to voluntarily accept their cost-effective responsibility or to transmit unbiased cost

1. This estimate can be found in Council on Environmental Quality, (1980, p. 179).

information to the control authority so it can make a cost-effective assignment. Plant managers have an incentive to accept as little control responsibility as possible in order to maintain or strengthen their competitive positions.

In this policy environment it is not surprising that the command-and-control allocation is not, and by itself could not become, cost effective. What may be surprising in light of the complexity of the task is that cost effectiveness is not an unreasonable objective for other approaches.

COST-EFFECTIVE PERMIT MARKETS

The reason emission reduction credits can result in a cost-effective allocation is quite straightforward. Plants have very different costs of controlling emissions. When credits are transferable, those plants that can control most cheaply find it in their interest to control a higher percentage of their emissions because they can sell the excess. Buyers for these reductions can be found whenever it is cheaper to buy emission reduction credits for use at a particular plant than to install more control equipment. Whenever an allocation of control responsibility is not cost effective, further opportunities for trade exist. When all such opportunities have been fully exploited, the allocation is cost effective.

Emissions trading solves the problems of information and incentive that are posed by command-and-control by allowing each participant to play that role it plays best. Regulators ensure that sources have the proper incentives by limiting the set of possible transfers to those consistent with meeting the pollution targets. By exploiting the flexibility inherent in emissions trading to lower their own costs, within the boundaries established by the control authority, individual sources lower the total costs incurred by all sources collectively. Self-interest in this case coincides with cost effectiveness.

Though these general principles hold regardless of the type of pollutant being regulated, some of the implementation details (such as the design of the emission reduction credit) depend rather crucially on the nature of the pollutant being regulated. Three different classes of pollutants are considered in this chapter. The definition of the cost-effective allocation as well as the emissions trading system designed to achieve that allocation varies among these pollutant classes. The characteristic that distinguishes any one of these pollutant classes from another is the relationship between individual source emissions and the pollution target.

Uniformly Mixed Assimilative Pollutants

The first, and, in a number of ways the simplest, class of pollutants to control is conventionally referred to as *uniformly mixed assimilative pollutants*. In the case of assimilative pollutants, the capacity of the environment to absorb them is sufficiently large, relative to their rate of emission, that the pollution level in any year is independent of the amount emitted in previous years. Simply put, assimilative pollutants do not accumulate over time.

In the case of uniformly mixed pollutants, the ambient concentration depends on the total amount of emissions, but not on the distribution of these emissions among the various sources. All distribution of the control responsibility within an airshed yielding the same total emissions will produce approximately the same effect on the pollution target. Volatile organic compounds provide one example of a type of pollution which fits this description. Their contribution to ozone, the pollutant of interest, is not thought to be sensitive to where they are emitted in the airshed.

Each of these characteristics (assimilation and uniform mixing) limits the complexity of a cost-effective permit system. The former allows the control process to ignore the difficult problem of pollutant accumulation, while the latter characteristic eliminates the need to worry about the location of the sources in designing the control policy—a significant advantage.

Symbolically, the relationship between source emissions and the pollution target for a uniformly mixed assimilative pollutant can be written as:

$$A = a + b \sum_{j=1}^{J} (\bar{e}_j - r_j) \tag{1}$$

where A is the steady-state level of pollution in a year, $\bar{e}_j$ is the steady-state emission rate of the jth source that would prevail if the source failed to control any pollution at all (hereafter referred to as the *uncontrolled emission rate*), r_j is the amount of emission reduction achieved by the jth source, J is the total number of sources to be regulated, and both "a" and "b" are parameters. Typically the "a" parameter is used to represent background pollution (from natural sources or sources, which for one reason or another, are not regulated), and the "b" parameter is simply a constant of proportionality.

Cost effectiveness in this context is defined as that allocation of emission levels among the J sources which meets the pollution target (designated $\bar{A}$) at minimum cost. Let $C_j(r_j)$ be the continuous cost function which represents the minimum cost to the source of achieving any level of emissions reduction r_j. Generally as r_j increases, the marginal cost of control can also be presumed to increase.

Mathematically the cost-effective allocation is the solution to the following minimization problem:

$$\min_{r_j} \sum_{j=1}^{J} C_j(r_j) \tag{2}$$

subject to

$$a + b \sum_{j=1}^{J} (\bar{e}_j - r_j) \leq \bar{A}$$

and

$$r_j \geq 0 \quad j = 1, \ldots, J$$

Thanks to some important theorems developed in the 1950s, it is possible to state the necessary and sufficient conditions (the Kuhn–Tucker conditions) for an allocation of control responsibility among the sources (a J-dimensional r-vector) to be cost effective[2]:

$$\frac{\partial C_j(r_j)}{\partial r_j} - \lambda b \geq 0 \qquad j = 1, \ldots, J \tag{3}$$

$$r_j \left[\frac{\partial C_j(r_j)}{\partial r_j} - \lambda b \right] = 0 \qquad j = 1, \ldots, J \tag{4}$$

$$a + b \sum_{j=1}^{J} (\bar{e}_j - r_j) \leq \bar{A} \tag{5}$$

$$\lambda \left[a + b \sum_{j=1}^{J} (\bar{e}_j - r_j) - \bar{A} \right] = 0 \tag{6}$$

$$r_j \geq 0; \; \lambda \geq 0 \qquad j = 1, \ldots, J \tag{7}$$

The λ variable is known as the Lagrange multiplier; it is introduced as a convenient means of solving the type of problem posed here. In this context it also has a simple economic interpretation—it is the amount of control cost that could be saved if the environmental quality constraint, $\bar{A}$, were relaxed by one unit. It is a measure of the marginal difficulty of meeting the standard $\bar{A}$.

These equations have a rather straightforward interpretation. In a cost-effective allocation of a uniformly mixed assimilative pollutant, the marginal cost of control for each source would be equal to the same constant (λb). This implies that the marginal cost of control would be the same for all sources. As we shall see, this is a highly significant finding.

Other properties of the cost-effective allocation can be extracted from these conditions. If, in a cost-effective allocation, the marginal cost of controlling the first unit of emission reduction for that source is higher

---

2. Kuhn and Tucker (1951).

than λb and, hence, higher than the marginal costs of all sources controlling non-zero amounts, that particular source is not assigned any control responsibility (that is, $r_j = 0$). It must also be the case that $\lambda > 0$ as long as some control is needed. The condition $\lambda = 0$ simply implies that the uncontrolled emissions satisfy the environmental constraint; no control is necessary.

The cost-effective emission reduction credit design for a uniformly mixed assimilative pollutant is called an *emission permit*. An emission permit is defined in terms of an allowable emissions rate (such as tons per year or pound per hour). To initiate an emission-permit system, the control authority must define the amount of emissions (N) which will be allowed. This is calculated as:

$$N = \sum_{j=1}^{J} (\bar{e}_j - r_j) = \frac{\bar{A} - a}{b} \tag{8}$$

since this is the level of allowed emissions which will cause the environmental quality standard to be met with equality.

For some pollutants both the "a" and "b" parameters change seasonally, implying that the number of permits should change seasonally as well. When the permit system reflects these seasonal effects, different marginal costs of control can prevail in different seasons.

Once these permits are issued, they would command a positive price as long as any control is needed to meet the target. Each source would attempt to acquire that number of permits consistent with minimizing its cost. Suppose that each source has some initial endowment of these permits (q_j^0). Across all sources this initial endowment must be equal to the number of allowable permits (that is, $\sum_{j=1}^{J} q_j^0 = N$) in order to ensure compliance with $\bar{A}$.

Faced with the need to choose a nonnegative level of control, the jth source's choice can be characterized as:

$$\min_{r_j} C_j(r_j) + P(\bar{e}_j - r_j - q_j^0) \tag{9}$$

where P is the price the source would pay for an acquired permit or receive for a permit sold to another source.

The solution for the set of all sources is:

$$\frac{\partial C(r_j)}{\partial r_j} - P \geq 0 \qquad\qquad j = 1, \ldots, J \tag{10}$$

$$r_j \left[\frac{\partial C_j(r_j)}{\partial r_j} - P \right] = 0 \qquad\qquad j = 1, \ldots, J \tag{11}$$

$$r_j \geq 0 \qquad\qquad j = 1, \ldots, J \tag{12}$$

This market solution can be compared with the cost-effective solution. According to equation (8), the number of permits issued is compatible with $\overline{A}$, so the environmental quality constraint would be satisfied. From the remaining equations it is clear that the permit system would yield the cost-effective allocation as long as $P = \lambda b$.

It turns out that the permit market would automatically yield this price. To see this, consider figure 1, which can serve the twin purpose of showing how the price is determined as well as demonstrating in a graphical way why permit markets are cost effective.

Figure 1 is drawn assuming that there are only two sources. For each source $\overline{e}_j = 15$ so the total uncontrolled emission rate is 30 units. The pollution target is assumed to be 15 units of allowed emissions, implying that the two sources together must reduce emissions by some 15 units if the target is to be met. The origin for the marginal cost of control for the first source (MC_1) in figure 1 is the left-hand axis and the origin for the marginal cost of control for the second source (MC_2) is the right-hand axis. Notice that the desired 15-unit reduction is achieved for every point on this graph. Drawn in this manner, the diagram represents all possible allocations of the 15-unit reduction between the two sources. The left-hand axis, for example, represents an allocation of the entire control responsibility to the second source, while the right-hand axis represents

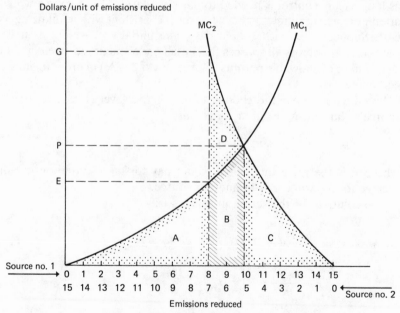

Figure 1. Cost effectiveness and the emission permit system

a situation in which the first source bears the entire responsibility. All points in between represent different degrees of shared responsibility.

It is easy to show that in a cost-effective allocation of the control responsibility between these two sources, the first source cleans up 10 units while the second source cleans up 5. The total variable control cost of this particular assignment of responsibility for the reduction is represented by the area $A + B + C$. Area $A + B$ is the cost of control for the first source while area C is the cost of control for the second. Any other allocation would result in a higher total control cost.

Figure 1 also demonstrates the important proposition derived earlier. *The costs of achieving a given reduction in emissions will be minimized if and only if the marginal costs of control are equalized across all emitters.* [3] Because the marginal cost curves cross at the cost-effective allocation, they must be equal at that point. Given the presumed shape of the curves, marginal control costs are not equal at any other allocation.

What this analysis implies for trading in emission reduction credits can also be seen in figure 1. Suppose that prior to any emissions trading the first source had to control 8 units. Since it has 15 units of uncontrolled emissions, this would mean it was allowed 7 units of emissions. Similarly, suppose the second source was assigned 7 units of reduction, meaning that in the absence of any trade it would be allowed to emit 8 units. Notice that both firms have an incentive to trade since the marginal cost of control for the second source (G) is substantially higher than that for the first (E). The second source would lower its cost as long as it could buy emission reduction credits from the first source at a price lower than G. The first source meanwhile would be better off as long as it could sell the credits for a price higher than E. Since G is greater than E, there are certainly grounds for trade.

A transfer of emission reduction credits would take place until the first source was controlling 10 units (2 more than originally) while the second source was controlling only 5 (2 less than originally). At this point the emission reduction credit price would equal P (since that is the value of a marginal unit of emissions to both sources) and neither source would have any further incentive to trade. Not only would the emission reduction credit market be in equilibrium, but the equilibrium would result in a cost-effective allocation of control responsibility.

As a result of the transfer of emission reduction credits equal to two units of emissions, the first source voluntarily controls more and the second source less. Allowing emission reduction credits to be traded

3. This is strictly true only if all sources are assigned some control responsibility in a cost-effective allocation. Two possible exceptions occur when no control is needed ($\lambda = 0$) or when one or more sources (say the jth) are so costly to control that they are assigned no control responsibility ($r_j = 0$).

results in a lower cost of compliance without any change in total emissions. (In this example, area D represents the amount saved by allowing emissions trading.) With emissions trading, the control authority can achieve a cost-effective allocation despite its lack of knowledge about control costs. It would merely define the total level of emission reduction, leaving the ultimate choice about control responsibility up to the sources who have the best information on the available control technologies.

Nonuniformly Mixed Assimilative Pollutants

A second and somewhat more complex class of pollutants involves a relationship between emissions and the pollution target for which the location of the sources is crucial. A number of important air and water pollutants such as total suspended particulates, sulfur dioxide, and biochemical oxygen demand fall within this classification.

For these pollutants, the policy target is specified in terms of a ceiling on the permissible ambient concentration of that pollutant measured at specific receptor locations.[4] Location is important because those concentrations are sensitive not only to the level of emissions, but to the degree of source clustering as well. Clustered sources would be more likely to trigger violations of the standard than dispersed sources with the same aggregate emission rate because, with clustering, the emissions would be concentrated in a smaller volume of air.

For this class of pollutants the emissions–environmental quality relationship can be written as:

$$A_i = \sum_{j=1}^{J} d_{ij} (\bar{e}_j - r_j) + a_i \qquad i = 1, \ldots, I \qquad (13)$$

where A_i is the concentration level measured at the ith receptor, a_i is the background pollution level at that receptor, d_{ij} is a transfer coefficient which translates emission increases or decreases by the jth source into changes in the concentration measured at the ith receptor, and I is the number of receptors. The transfer coefficient expresses a steady-state relationship and takes into account such factors as average wind velocity and direction, the locations of sources and receptors, as well as source stack heights.

The cost-effective allocation of a nonuniformly mixed assimilative pollutant is that allocation which minimizes the cost of pollution control

4. The law specifies that these ceilings are to be met *everywhere* and not merely at specified monitored locations. As a practical matter, however, studies have shown that a relatively few monitors (say 9 or 10) can effectively cover a particular region of interest. See Ludwig, Savitz, and Valdes (1983).

subject to the constraint that the I-dimensional vector of predetermined concentration ceilings $\overline{A}$ is met at all receptors. Symbolically

$$\min_{r_j} \sum_{j=1}^{J} C_j(r_j) \tag{14}$$

subject to $\overline{A}_i \geq a + \sum_{j=1}^{J} d_{ij} (\overline{e}_j - r_j)$ $i = 1, \ldots, I$

and $r_j \geq 0$ $j = 1, \ldots, J$

Once again we can use the knowledge that a cost-effective allocation must satisfy the Kuhn–Tucker conditions to learn something about the nature of that allocation. For the problem defined in (14) the Kuhn–Tucker conditions are:

$$\frac{\partial C_j(r_j)}{\partial r_j} - \sum_{i=1}^{I} d_{ij} \lambda_i \geq 0 \qquad j = 1, \ldots, J \tag{15}$$

$$r_j \left[\frac{\partial C_j(r_j)}{\partial r_j} - \sum_{i=1}^{I} d_{ij} \lambda_i \right] = 0 \qquad j = 1, \ldots, J \tag{16}$$

$$\overline{A}_i \geq a + \sum_{j=1}^{J} d_{ij} (\overline{e}_j - r_j) \qquad i = 1, \ldots, I \tag{17}$$

$$\lambda_j \left[\overline{A}_i - a + \sum_{j=1}^{J} d_{ij} (\overline{e}_j - r_j) \right] = 0 \qquad i = 1, \ldots, I \tag{18}$$

$$r_j \geq 0; \quad i \geq 0 \qquad \begin{matrix} j = 1, \ldots, J \\ i = 1, \ldots, I \end{matrix} \tag{19}$$

Equation (15) states that for nonuniformly mixed assimilative pollutants in a cost-effective allocation, each source should equate its marginal cost of emission reduction with a weighted average of the marginal cost of concentration reductions (λ_i) at each affected receptor. The weights are the transfer coefficients associated with each receptor. If the cost-effective pollutant concentration were lower than the ceiling at any receptor, equation (18) implies that the λ_i associated with that receptor would be zero. This is referred to as a *nonbinding receptor*. For any binding receptor, the associated λ_i would be positive. Notice that for this class of pollutants it is not the marginal costs of emission reduction that are equalized across sources in a cost-effective allocation (as was the case for uniformly mixed assimilative pollutants), it is the marginal costs of concentration reduction at each receptor location that are equalized.

Seasonality in nonuniformly mixed assimilative pollutants not only affects the amount of allowable emissions (a characteristic it shares with uniformly mixed pollutants), but it also affects transfer coefficients. Wind velocity and direction patterns frequently have a distinct seasonal component. Because of these seasonal effects, a cost-effective allocation frequently involves not only different seasonal levels of regional control, but also different seasonal allocations of the emission reduction responsibility among sources.

The permit system designed to yield a cost-effective allocation of the control responsibility for nonuniformly mixed assimilative pollutants is called an *ambient permit* system. Because this system involves a separate permit market associated with each receptor, each source would have to procure sufficient permits in each of the I markets to legitimize its emission rate.

Faced with the need to acquire permits from I markets, the source's decision on the level of control can be characterized as:

$$\min_{r_j} C_j(r_j) + \sum_{i=1}^{I} P_i \left[d_{ij}(\bar{e}_j - r_j) - q_{ij}^0 \right] \qquad (20)$$

where P_i is the price that prevails in the ith permit market and q_{ij}^0 is the pretrade amount of concentration units at the ith receptor allowed to the jth source. The allocation which minimizes this cost is:

$$\frac{\partial C_j(r_j)}{\partial r_j} - \sum_{i=1}^{I} P_i d_{ij} \geq 0 \qquad j = 1, \ldots, J \qquad (21)$$

$$r_j \left[\frac{\partial C(r_j)}{\partial r_j} - \sum_{i=1}^{I} P_i d_{ij} \right] = 0 \qquad j = 1, \ldots, J \qquad (22)$$

$$r_j \geq 0 \qquad j = 1, \ldots, J \qquad (23)$$

A comparison of equations (21)–(23) with equations (15)–(19) reveals $P_i = \lambda_i$ to be a sufficient condition for this permit system to produce the cost-effective allocation. As long as the control authority issues the appropriate number of permits for each receptor, the equivalence of supply and demand would ensure that $P_i = \lambda_i$ in each market.

Whereas emission permits are defined in terms of allowable emission rates, ambient permits are defined in terms of units of concentration at the receptor locations. The amount of allowed concentration at each receptor is determined by subtracting the background concentration (a_i) from the concentration permitted by the ambient standard ($\bar{A}_i$).

The denominations of these permits can be dictated by convenience. If the control authority wants a large number of permits to accommodate small sources, it could issue 1,000 permits, each denominated in units of 1/1,000 of the allowed concentration. If a smaller number of permits were desired, it could issue 100 permits, each worth 1/100 of the allowed concentration increase.

To explain how and why an appropriately designed emissions trading program would work in this context, it will simplify matters to consider initially a simple numerical example involving two sources and a single receptor (table 3). Once the single receptor case is understood, it is rather easy to deal with the added complexity of multiple receptors.

Assume that the two sources have the same marginal cost curves for cleaning up the emissions, as reflected by the identical first two corresponding columns of the table for each of the two sources. The main difference between the two sources is their location in relation to the receptor. Because the first source is assumed to be closer to the receptor, it has a larger transfer coefficient than the second (1.0 as opposed to 0.5).

Suppose the objective is to meet a given concentration target (3.0 units) at minimum cost and that with no control each source would emit 7 units of emission, resulting in 10.5 units of pollution concentration at the receptor. The relationship between emissions and concentration at the receptor for each source is given by column 3 while column 4 records

Table 3. Cost Effectiveness for Nonuniformly Mixed Assimilative Pollutants: A Hypothetical Example

Source 1 ($d_{11} = 1.0$)			
Emissions units reduced	Marginal cost of emission reduction (dollars per unit)	Concentration units reduced[a]	Marginal cost of concentration reduction[b] (dollars per unit)
1	1	1.0	1
2	2	2.0	2
3	3	3.0	3
4	4	4.0	4
5	5	5.0	5
6	6	6.0	6
7	7	7.0	7
Source 2 ($d_{12} = 0.5$)			
Emissions units reduced	Marginal cost of emission reduction (dollars per unit)	Concentration units reduced[a]	Marginal cost of concentration reduction[b] (dollars per unit)
1	1	0.5	2
2	2	1.0	4
3	3	1.5	6
4	4	2.0	8
5	5	2.5	10
6	6	3.0	12
7	7	3.5	14

[a] Computed by multiplying the emission reduction in column 1 by the transfer coefficient (d_{ij}).
[b] Computed by dividing the marginal cost of emission reduction (column 2) by the transfer coefficient (d_{ij}).

the marginal cost of each unit of concentration reduced. The former is merely the emission reduction times the transfer coefficient while the latter is the marginal cost of the emission reduction divided by the transfer coefficient (which translates the marginal cost of *emission* reduction into a marginal cost of *concentration* reduction).

Suppose as part of a command-and-control approach to this pollution the control authority required the first source to clean up 4.0 units of emissions and the second source to clean up 7.0 units. This would meet the standard $[(4)(1.0) + (7.0)(0.5) = 7.5]$, but would not be cost effective.

The control costs have to be calculated to verify this. Costs can be calculated for each source control by adding up the marginal costs of emission control associated with each control unit. The control cost for the first source would be $\$10^5$ while the control cost for the second source would be $28, producing a total cost for the two sources of $38.

If an emissions trading program were established, the sources would be free to trade concentration units. In particular, assume that the first source decided to clean up 6 units of emissions rather than its command-and-control requirement of 4 units, selling the extra 2 units to the second source. By acquiring 2 concentration units from the first source, the second source could control 3 units of emissions rather than 7. It would gain the right to emit two more concentration units, which translates into 4 more units of emissions. The total concentration reduction (7.5) would be the same as that for the command-and-control policy, but the control cost of the emission trading allocation would be lower. The emission trading control costs for the first source would be $21, while those for the second source would be $6. The control costs for both sources taken together ($27) would be lower than those for the command-and-control policy ($38).

Allowing concentration trades reduces total control costs. The first source voluntarily undertakes more control because he or she can sell the concentration reduction credits for more than it costs to produce them. In this example the credits would sell for $6 apiece since that is the price at which the marginal costs of concentration reduction are equalized for the two sources. At that price the first source can sell for $12 what cost $11 to produce. The second source would pay $12 for the two credits, but would save $22 in control costs. The trade is in the interest of both parties. Furthermore, since the marginal cost of reducing the concentration to the desired level is equal for both sources, this trading equilibrium is cost effective.

5. The marginal costs of the first, second, and third unit of reduction for the first source are, respectively, $1, $2, and $3, for a 3-unit control cost of $6. With a marginal cost of $4, the 4-unit control cost is $10. The calculations for the second source are similar.

The extension of this system to the many-receptor case requires that a separate concentration reduction credit be created *for each receptor*. The price prevailing in each of these markets would reflect the difficulty of meeting the ambient standard at that receptor. All other things being equal, concentration reduction credits associated with receptors in heavily congested areas could be expected to sustain higher prices than those affected by a relatively few emitters.

In multiple-receptor, ambient-permit markets, marginal *emission* control costs would vary across sources for two reasons. First, for any particular receptor, equal marginal costs of concentration reduction do not imply equal marginal costs of emission reduction. Sources with larger transfer coefficients would face higher marginal costs of emission reduction. Second, some receptors would require more control than others, and sources having larger transfer coefficients associated with those receptors would face higher relative marginal costs of emission reduction.

This variation of costs with location sends a signal to new sources who are deciding where to locate. The high control costs associated with heavily polluted areas provide an incentive for heavy emitters to locate elsewhere. Even though pollution control expenditures are only part of the costs a firm considers when deciding where to locate, they are a factor.

The many-receptors case is the appropriate one for discussing regional pollutants such as acid rain. Because these pollutants tend to be transported long distances by the prevailing winds, receptors rather far from the source would be affected by such emissions. The ambient permit system handles this geographic interdependency by requiring sources to purchase permits from all affected receptors, remote as well as local. Remote receptor permits would also be defined in terms of allowable concentration increases, with transfer coefficients used to relate emissions to concentration increases. The acid rain case is therefore conceptually no different than the general nonuniformly mixed case; only the proximity of the receptors is at issue.

The similarities between local and regional pollution problems should not be allowed to obscure the differences, however. The transfer coefficients that relate emissions of sulfur and nitrogen oxides in one state or country to concentrations in another state or country are measured with considerably less reliability than transfer coefficients measured for a particular, relatively small, geographic area such as a city.[6] Thus while the principles are the same, implementing those principles is much more difficult for regional than for local pollutants.

6. A review of the models used to simulate the long-range transport of pollutants can be found in Johnson (1983).

Uniformly Mixed Accumulative Pollutants

The final class of pollutants we consider contains substances which accumulate in the environment because their rate of injection exceeds the assimilative capacity. In this section we assume that location does not matter because for one important pollutant, chlorofluorocarbons, that is the most reasonable specification and because this assumption reduces the notational complexity. For some other accumulative pollutants (such as lead), location would matter.

For a uniformly mixed, accumulative pollutant, a common form of the pollution target-emission rate relationship can be described as:

$$A_t = a + \sum_{j=1}^{J} \sum_{k=1}^{t} (\overline{e}_{jk} - r_{jk}) \tag{24}$$

where the pollution level in any year t is the simple addition of the initial pollution level (a) plus all emissions in the intervening years. This formulation implicitly assumes no assimilative capacity and a one-to-one correspondence between a 1-unit increase in emissions and a 1-unit increase in pollution.

Suppose that the policy objective was to ensure that A_t would never be higher than some ceiling $\overline{A}$. Since the pollution-emission relationship specified in equation (24) allows no means of reducing the size of the stock of accumulated pollutants, once the ceiling is reached, no further emissions are permitted. The question of interest for this class of pollutants is not only how the cost-effective control responsibility is allocated among sources, but how it is allocated over time as well.

The cost-effective allocation in this context is the one which has the lowest associated present value of control costs among all those allocations which satisfy the pollution constraint. The decision problem can be symbolically stated as:

$$\min_{r_{jt}} \sum_{t=1}^{T} \sum_{j=1}^{J} \frac{C_j(r_{jt})}{(1 + \rho)^{t-1}} \tag{25}$$

subject to
$$\overline{A} \geq a + \sum_{j=1}^{J} \sum_{t=1}^{T} (\overline{e}_j - r_{jt})$$

and
$$r_{jt} \geq 0 \qquad\qquad t = 1, \ldots, T$$

where ρ is the interest rate used to translate future costs into their present-value equivalents and T is the length of the planning horizon. Relying once again on the Kuhn–Tucker conditions, it is possible to derive the conditions any cost-effective allocation will satisfy:

$$\frac{\partial C_j(r_{jt})}{\partial r_{jt}} \frac{1}{(1 + \rho)^{t-1}} - \lambda \geq 0 \qquad \begin{array}{l} j = 1, \ldots, J \\[4pt] t = 1, \ldots, T \end{array} \tag{26}$$

$$r_{jt} \left[\frac{\partial C_j(r_{jt})}{\partial r_{jt}} \frac{1}{(1+\rho)^{t-1}} - \lambda \right] = 0 \qquad \begin{aligned} j &= 1, \ldots, J \\ t &= 1, \ldots, T \end{aligned} \tag{27}$$

$$\overline{A} - a - \sum_{j=1}^{J} \sum_{t=1}^{T} (\overline{e}_j - r_{jt}) \geq 0 \tag{28}$$

$$\lambda \left[\overline{A} - a - \sum_{j=1}^{J} \sum_{t=1}^{T} (\overline{e}_j - r_{jt}) \right] = 0 \tag{29}$$

$$r_{jt} \geq 0; \quad \lambda \geq 0 \qquad \begin{aligned} j &= 1, \ldots, J \\ t &= 1, \ldots, T \end{aligned} \tag{30}$$

To the nonmathematical reader this is probably the most intimidating set of equations yet, but, as we found with the others, an intuitive understanding of what they convey is possible. In those years when some, but less than complete, control is being exercised, the cost-effective control levels will satisfy:

$$\frac{\partial C_j(r_{jt})}{\partial r_{jt}} = (1 + \rho) \frac{\partial C_j(r_{jt-1})}{\partial r_{jt-1}} \qquad \begin{aligned} j &= 1, \ldots, J \\ t &= 1, \ldots, T \end{aligned} \tag{31}$$

This implies that in a cost-effective allocation, marginal pollution control costs rise over time at rate ρ and the amount emitted declines over time. In each time period the marginal costs of control are equalized across all sources.[7]

If T is long enough, as t increases eventually a year is reached in which the ambient constraint (equation 28) becomes binding and allowable emissions cease from then on. The marginal control costs at that point are those associated with complete control; they are no longer necessarily equalized.

The permit system yielding this cost-effective allocation is called a *cumulative emission permit* system. The permits themselves do not have a time dimension; the holder has complete freedom when to emit. However, in contrast to the two previously discussed permit systems, cumulative emission permits do not regulate emission rates; they limit total emissions (tons rather than tons per year). Once this total emission ceiling is reached, the permit allows no more emissions. In this market the permits are an exhaustible resource; once used, they are withdrawn from circulation.

7. This analysis presumes smooth cost functions. In practice it may be necessary to purchase emissions control in large discrete increments. In this case the source would make a few large investments, producing more control in the earlier years than expected on the basis of smooth cost functions. Eventually even in this case total control would be necessary; only the path by which this outcome would be reached would differ.

Defining the appropriate number of permits is the first step that must be taken by the control authority in order to establish this market. The appropriate number is dictated by the environmental quality constraint; the amount of allowable emissions equals the pollution target minus the pollution already in the environment. Once these permits are issued, a market price would be established. The supply of unused permits diminishes over time (as some are used in the earlier years), while the demand for them increases. Prices rise to bring demand and supply into balance in each year.

Due to the demand and supply patterns, the rate of increase in permit prices would be equal to p, the rate of interest. This implies that sources respond to these rising prices by choosing that level of emission control that equates marginal control cost and permit price. Since prices would rise at rate p, the rate of increase in marginal control costs would also be p, precisely what is required for the allocation to be cost effective.

The notion that sources would conserve some permits for later years may not be obvious. What dictates conservation in this model? Any source that uses its permits too early forgoes the substantial value of those permits in the future as reflected in the higher prices. Myopic behavior raises the firm's costs unnecessarily and that is inconsistent with our assumption that sources minimize their costs.

OTHER PROPERTIES OF TRANSFERABLE PERMIT SYSTEMS

The Information Burden

This chapter began with a description of the rather large information burden placed on a control authority if it is to assign emission standards in a cost-effective manner. Not only must the authority know all possible means of reducing emissions for all types of sources, it must also know the marginal costs of control associated with various levels of control for each of the means. Furthermore, it must continually be aware of changes in the menu of control possibilities to ensure that the most cost-effective means are continually being used, not used merely when the source is first subject to control.

Are the information requirements for a transferable permit system more realistic? Because permit systems tend to shift the information burden to those most able to derive and to act on this information (the sources), the burden of control is both diminished and realistic.

Of the three types of systems, the burden is heaviest for an ambient permit system. In this type of permit market, air dispersion models must be used by the control authority to derive the transfer coefficients. Even this need is not unique to permit systems since air dispersion models were already in use by the EPA in air pollution policy prior to the current reforms. Though their validity is less than ideal, their use has successfully withstood a number of legal challenges.[8]

Previous court successes do not automatically imply future successes. In the past, ambient modeling has been used primarily to decide whether a particular emission rate would trigger a violation of an ambient standard. This "yes or no" choice is a qualitatively different use of air quality modeling than deriving a complete set of source-receptor transfer coefficients.

This dispersion modeling burden is not a trivial one, but it must be kept in perspective. It is a substantially easier burden for the control authority to bear than that associated with assigning emission standards. In contrast to the information requirements of the command-and-control system, an ambient permit system does not require any information on the cost of control, the most difficult information burden to bear.

It is easy to imagine improvements in dispersion modeling as the system matures, but, because of the dynamic nature of control technology, it is difficult to imagine comparable gains in collecting cost information. Dispersion modeling has the distinct advantage of attempting to capture physical processes which over the long term are stable while cost functions are by their very nature continuously changing. Because it is tougher to hit a moving target, it is less reasonable to expect the control authority to have up-to-date information on control costs than on pollutant flows.

The emissions permit and cumulative emission permit systems have even smaller information burdens than the ambient permit system. They can be initiated once the proper number of permits has been established. Establishing that number requires no information on control costs or on transfer coefficients. Because they are driven solely by the politically predetermined pollution target which, by the nature of the pollutant, can be rather easily translated into an emissions or emissions rate target, the burden is almost a trivial one. In any case, it poses no barrier to the adoption of these types of permit systems.

The Speed of Compliance

The second stated objective for the reform package was for it to hasten compliance. Why is it reasonable to believe that introducing transferability into the permit process would hasten compliance? On the surface this expectation may appear paradoxical since the emissions standard

8. See, for example, Pierce and Gutfreund (1975) and Case (1982).

system involves sources being told what to do and how to do it. What could be quicker?

The paradox is resolved when the behavioral responses of the sources to emission standards are compared with those for any of the above permit systems. In principle, permit transferability not only encourages sources to adopt existing pollution control techniques more rapidly, it also encourages the continued development of new, more effective techniques.

Under a command-and-control system, a source has two options: (1) it can adopt the technology or (2) it can fight its assigned standard. Because this system is not cost effective, some sources inevitably face quite high compliance costs. For these sources, the lowest cost solution is frequently to challenge its assigned standard. Common legal grounds include the assertion that the source cannot afford to comply if it is to stay in business or that the standard is not technologically feasible (for example, it is based upon a relatively unproven technique in actual commercial operations). These challenges frequently succeed.

With a transferable permit system in place, the circumstances are changed. As long as emission reduction credits are available to high-cost sources, their purchase is frequently a cheaper way to meet legal requirements than litigation. The source's advantage in litigation (the ability to reduce the cost of compliance) is smaller and the likelihood of receiving a favorable verdict is reduced when permits are available. The court is less likely to be persuaded by the source's arguments on affordability when relatively cheap alternatives exist. Furthermore, the technological infeasibility arguments lose force when several means of complying are available. Whereas under a command-and-control system the emission standards are based on a particular means of control, opening the feasibility of that control technique to legal challenge, permit systems leave the choice of techniques to the source. Sources would then have to prove no feasible technique exists, a rather more difficult legal burden to bear. These arguments suggest that a permit system could be expected to hasten compliance, compared with a command-and-control system, because sources would have more existing control options at their disposal.

A second dimension of this argument suggests that permits not only allow more flexibility in choosing among *existing* control techniques, but they stimulate the development of *new* techniques as well. Consider figure 2. Suppose the source is currently using a technology with a marginal cost of control MC^0, but it is contemplating switching to one that would result in a lower marginal control cost, MC^1. Should it switch? If the cost savings of the switch were larger than the purchase price of the new technology, the source would save money by switching.

The advantage of switching (the cost savings) would be larger with a permit system than it would be if the source were facing a nontrans-

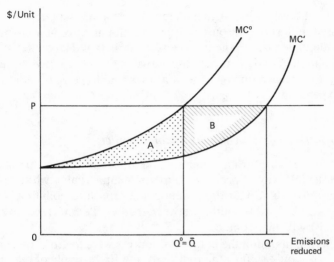

Figure 2. Cost savings from technical change: Permits vs. emission standards

ferable emission standard. As drawn, prior to the availability of the new control technique the source would be controlling Q^0 emissions; this same control level would be achieved with either a permit price P or an emission standard $\bar{Q}$. Because sources can sell emission reduction credits for P, the cost savings brought about by a switch to the new technique would be $A + B$. Area A represents the lower cost of meeting the previous standard (Q^0) while area B is the profit earned from selling $Q^1 - Q^0$ emission reduction credits. With the command-and-control approach there would be no further savings from emission reductions beyond Q^0 since further reductions cannot be sold; the savings from the switch would be equal to A. Since the equipment costs the same in either case, the greater savings with the emission reduction credit increases the likelihood that the switch would take place. If the switch takes place, more emissions are controlled by that source and the costs of compliance for both trading sources are reduced.

The essential point is that with transferable permits sources would have a continual interest in lower cost means of reducing their emissions, since cost reductions would improve their competitive position. With nontransferable emission standards, the development of new control techniques would be of less consequence. Once they have met the standard, sources would have less to gain from further emission reductions.

These incentives affect the potential suppliers of new technologies as well. When permits are transferable, the demand for new technologies is higher. Faced with a higher demand potential, suppliers are more interested in devoting resources to the development of new techniques. Higher demand means potentially larger markets and larger sales. In the face of both higher demand and higher willingness to supply, more rapid adoption of new technologies with a transferable permit seems a plausible expectation.

The Growth–Environmental Protection Conflict

One of the reasons for implementing the offset policy was to introduce a more flexible way of protecting environmental quality while allowing economic growth. Under the traditional approach to pollution control, there was an inevitable conflict between the two. Permits serve to reduce the intensity of that conflict.

In a permit market if the number of sources were to grow, the demand for permits would shift to the right. Given a fixed supply of permits the price would rise, as would the control costs, but the amount of emissions (or pollution concentrations in the case of the ambient permit system) would remain the same.

Contrast this with what happens in a command-and-control approach. In the absence of additional action by the control authority, growth in the number of sources would not affect the emission standards imposed on each source. Existing sources would control only what they would control in the absence of growth and, therefore, the arrival of new sources would cause a deterioration of air quality in the region.

To prevent the further deterioration of air that was already worse than required by the standards, control authorities were forced into a situation where they had to either stop growth or decide on new, more stringent control responsibility assignments for existing sources. Neither of these alternatives was politically palatable.

Transferable permits, as embodied in the offset policy, allow a more flexible approach. Sources are free to enter a region provided that they do not make the air quality worse. Growth is allowed as long as it does not interfere with protecting air quality. Because they can sell their permits, existing sources are led to increase emission control voluntarily.

All parties potentially have something to gain from this flexibility. Control authorities are relieved of the need to stop growth or to decide how much additional control each existing source should bear. Existing sources are compensated for further reductions under the permit system; with more stringent emission standards they would bear the financial burden themselves. New sources have an option to move into regions which under the command-and-control policy would be closed to future growth.

SUMMARY

• The traditional command-and-control approach to air pollution control imposes a large information burden on control authorities. Because the magnitude of this burden exceeds their ability to respond, control authorities have promulgated emission standards that are generally neither cost effective nor capable of promoting rapid compliance.

• Cost-effective emissions trading systems can be designed, but the design depends crucially on the nature of the pollutant being regulated. Three common pollutant classes implying very different designs are: (1) uniformly mixed, assimilative pollutants, (2) nonuniformly mixed, assimilative pollutants, and (3) uniformly mixed, accumulative pollutants.

• A cost-effective allocation of uniformly mixed, assimilative pollutants could be achieved by an emissions permit system. For any geographic area this system allows ton-for-ton trades among any sources in the airshed. The cost-effective emission reduction credits would be defined in terms of an allowable emission rate. Total cumulative emissions would not be controlled.

• For nonuniformly mixed assimilative pollutants, an ambient permit system would yield the cost-effective allocation of control responsibility. For each airshed this system requires considering the spatial effects of trades on each receptor site. The cost-effective emission reduction credits in this approach would be defined in terms of allowable concentration increases at specific receptor locations. There would be separate credits associated with each receptor which could be banked and sold independently. Total cumulative emissions would not be controlled with this system either.

• Uniformly mixed accumulative pollutants can be cost effectively controlled using a cumulative emission permit system. A single emission reduction credit, defined in terms of emissions (tons), not emission rates (tons per year), would suffice. Total cumulative emissions would be controlled by this approach.

• The regulatory information burden necessary to achieve a cost-effective allocation of the control responsibility is smaller with emissions trading systems than with the command-and-control approach. Of the three considered permit approaches, ambient permit systems impose a higher burden than the other two permit systems since they require the use of air dispersion models to derive the transfer coefficients which govern the allowable transfers among sources.

• Compared with the command-and-control approach, theory suggests that emissions trading programs can be expected to encourage more rapid compliance by allowing a more complete exploitation of existing control techniques and by providing more encouragement for the development of new techniques.

• By providing a more flexible approach to the conflict between economic growth and environmental protection, emissions trading approaches reduce the intensity of the conflict. Because they allow growth as long as the ambient quality standards are not violated in the process, they provide a mechanism by which new sources can seek additional emission reductions from existing sources, compensating them for any transfers successfully negotiated.

REFERENCES

Baumol, William J., and Wallace E. Oates. 1975. *The Theory of Environmental Policy* (Englewood Cliffs, N.J., Prentice-Hall).

Case, Charles D. 1982. "Problems of Judicial Review Arising from the Use of Computer Models and Other Quantitative Methodologies in Environmental Decisionmaking," *Boston College Environmental Affairs Law Review* vol. 10, no. 2, pp. 251–363.

Council on Environmental Quality. 1980. *Environmental Quality—1980* (Washington, D.C., U.S. Government Printing Office).

Johnson, Warren B. 1983. "Interregional Exchanges of Air Pollution: Model Types and Applications," *Journal of the Air Pollution Control Association* vol. 33, no. 6 (June) pp. 563–574.

Kuhn, H. W., and A. W. Tucker. 1951. "Nonlinear Programming," in J. Neyman, ed., *Proceedings of the Second Berkeley Symposium on Mathematical Statistics and Probability* (Berkeley, Calif., University of California Press) pp. 481–492.

Ludwig, F. L., H. S. Savitz, and A. Valdes. 1983. "How Many Stations Are Required to Estimate the Design Value and the Expected Number of Exceedances of the Ozone Standard in an Urban Area," *Journal of the Air Pollution Control Association* vol. 33, no. 10 (October) pp. 963–967.

Pierce, D. F., and P. D. Gutfreund. 1975. "Evidentiary Aspects of Air Dispersion Modeling and Air Quality Measurements in Environmental Litigation and Administrative Proceedings," *Federation of Insurance Council Quarterly* vol. 25 (Spring) pp. 341–353.

ADDITIONAL READING

Bohm, Peter, and Clifford S. Russell. "Comparative Analysis of Alternative Policy Instruments," in Allen V. Kneese and James L. Sweeney, eds.,

Handbook of Natural Resource and Energy Economics (Amsterdam, North-Holland) forthcoming.

Montgomery, W. David. "Markets in Licenses and Efficient Pollution Control Programs," *Journal of Economic Theory* vol. 5, no. 3 (December 1972) pp. 395–418.

Tietenberg, Thomas H. "Controlling Pollution by Price and Standards Systems," *Swedish Journal of Economics* vol. 75, no. 2 (June 1973) pp. 193–203.

Wenders, John T. "Methods of Pollution Control and the Rate of Change in Pollution Abatement Technology," *Water Resources Research* vol. 11, no. 3 (June 1975) pp. 393–396.

3 / The Potential for Cost Savings

The theory in the preceding chapter provides a convincing basis for believing that the traditional command-and-control approach is not and cannot become cost effective. The amount of information required by the regulatory authority if it is to establish a set of cost-effective emission standards is so high as to preclude a cost-effective outcome. In contrast, because the information requirements for initiating a transferable permit system are lower, cost-effective permit systems are, in theory at least, a distinct possibility.

As interesting as these theoretical results are, they provide an incomplete guide to regulatory reform. Because any change in policy has its own set of costs, it is difficult to overcome the inertia of the status quo. New grounds for legal challenge are exposed. Bureaucratic staffs trained in one set of procedures must learn new ones. The comfort of familiarity is lost to both regulators and sources.

In order to overcome this inertia, successful reforms usually involve only a small departure from the traditional approach in order to hold down the costs of change as perceived by the participants. The EPA emissions trading program satisfied this condition because it built upon the traditional approach. Emission reduction credits are a complement to, not a replacement for, the traditional approach.

Successful reforms must also promise benefits substantial enough to outweigh any frictional costs of moving away from a traditional approach. The theoretical models explicated in chapter 2 have made it clear that the benefits to be realized from this reform (in the form of reduced compliance costs) are positive, but these models are incapable of rendering any assistance on the important question of whether or not the benefits are substantial. Answering that question requires an appeal to a different kind of evidence.

This chapter surveys two such kinds of evidence: (1) computer simulations comparing least-cost to command-and-control allocations, and (2) actual offset, netting, bubble, and banking transactions. Our objective is not only to discover whether the potential benefits are substantial, but also to identify from a comparison of these studies those regional characteristics that determine the size of these benefits. The reliability of these estimates as well as their magnitudes will be explicitly considered.

THE NATURE OF THE EVIDENCE

Simulation Studies

One way to gain information on the potential size of the benefits to be derived from a cost-effective transferable permit system prior to its full implementation is to compare the command-and-control allocation of control responsibility with a least-cost allocation in the context of a computer simulation model. The models used to perform these simulations depend heavily on the theoretical work covered in chapter 2. For each type of pollutant the Kuhn–Tucker conditions are used as the basis for algorithms designed to find the least-cost allocation of control responsibility for the specific pollutants and geographic areas being investigated.[1]

The simulations covered in this chapter encompass all three pollutant types defined in chapter 2, though the majority of studies deal with nonuniformly mixed assimilative pollutants. The coverage within that class of pollutants is rather extensive, involving a number of different substances as well as a variety of geographic settings. To facilitate comparisons, the costs associated with the command-and-control allocation are calculated for each study, along with the control costs associated with the least-cost allocation.

The chief virtue of these simulation models is that they allow the analyst the opportunity to examine counterfactual situations. As will be clear in subsequent sections of this book, there are a number of constraints operating on the emissions trading program. These arise from the statutes, from court decisions, or simply from the quest for

1. Specifically, the computer finds the source-specific emission reductions which solve equation sets (3)–(7) for uniformly mixed assimilative pollutants, equation sets (15)–(19) for nonuniformly mixed assimilative pollutants, or equation sets (26)–(30) for uniformly mixed accumulative pollutants.

bureaucratic convenience. Whatever the source, it is useful to know how seriously they jeopardize the degree to which the reforms can achieve the objectives established for them. Actual experience with emissions trading is not much help in this regard since the constraints usually apply to every transaction. Discovering the unique effects of any particular constraint requires a comparison of the allocations with and without the constraint. In the absence of controlled experiments, this comparison is possible only with simulation models.

The ability to obtain answers to "What if . . . ?" questions is a substantial virtue, but simulation models do have their drawbacks. Perhaps the chief drawback is that they deal with an idealized situation. Rather than simulating the actual workings of permit markets, these simulations find the least-cost allocation for meeting a particular standard. Relying on the theory presented in chapter 2 showing the cost effectiveness of permit markets, analysts use these allocations as if they were permit market allocations. Though this equivalence is valid for perfectly functioning markets, it is not necessarily valid for less-than-perfect markets. We return to this point later in the chapter.

Studies of Emissions Trading Transactions

A second source of information for our analysis is a by-product of EPA's responsibility for overseeing the emissions trading program. Since many of the emissions trading transactions require EPA approval, EPA has used this opportunity to accumulate a file on them. Many of the entries contain information on the cost savings resulting from approval of the transaction, the name and location of the purchaser, and the effect of the transaction on emissions. Though no systematic analysis of these data has been published, for our purposes we can derive quite a bit of useful information from the raw data.

Unfortunately the sample is a far from complete inventory of all aspects of emissions trading transactions. Many transactions that are in the file are missing key pieces of information, such as the magnitude of the realized cost savings. Because sources are not required to supply this information to gain approval of an emissions trading transaction, many chose not to provide it. In addition, many consummated transactions never enter the file because they do not require EPA approval. Both netting and offset transactions, for example, require only state approval. Furthermore, since EPA is delegating more approval authority to the states as they develop satisfactory criteria and procedures, even less EPA control over the information will be exercised in the future.

Incompleteness is not the only weakness of the transaction data. Since the estimates of cost savings and emission reductions are drawn from the applications, there are also reasons for being skeptical about the quality of the reported data. To maximize the chances that their application may be approved, sources have an incentive to overstate cost savings and emission reductions. Since the procedures for verifying these numbers are far from foolproof, the numbers have to be taken with a grain of salt.

Though simulation studies and transaction data each exhibit some weaknesses in isolation, combined they complement each other to a reassuringly high degree. The weaknesses of the simulation studies are countered by the strengths of the transaction data and vice versa. Whereas the simulation studies deal with idealized circumstances, the transactions data reflect real situations. Whereas the transaction data are incomplete and partial, the simulation studies are more global, covering the most significant sources and potential transactions in each region studied. Though far from ideal, together these data provide a respectable basis for evaluating the cost effectiveness of this reform package.

THE MAGNITUDE OF POTENTIAL COST SAVINGS

The Evidence

From the theory in chapter 2 we know that, in principle, appropriately designed transferable permit systems are capable of achieving a cost-effective allocation. In practice, the savings registered by emissions trading programs will be determined not only by the *potential cost savings,* which is measured as the deviation of the cost of the command-and-control allocation from the lowest possible cost of achieving the same pollution target, but also by the degree to which the costs resulting from actual emissions trading programs approximate the least-cost solution.

The magnitude of potential cost savings depends on many local circumstances, such as prevailing meteorology, the locational configurations of sources, stack heights, and how costs vary with the amount controlled. Several simulation models have now been constructed which integrate these factors for specific pollutants in specific airsheds (table 4). None of these studies considers the effects of seasonality, assuming instead that a single allowed emissions level and a single transfer coefficient matrix are appropriate.

Since for a variety of reasons the estimated costs cannot be directly compared across studies, the potential cost savings are presented as the ratio of command-and-control costs to the lowest cost of meeting the same objective. A ratio equal to 1.0 implies that the command-and-control allocation is cost effective and, therefore, the potential cost savings are zero. A ratio greater than 1.0 implies positive potential cost savings. When 1.0 is subtracted from this ratio and the remainder is multiplied by 100, the result can be interpreted as the percentage

Table 4. Empirical Studies of Air Pollution Control

Study and year	Pollutants covered	Geographic area	CAC benchmark	Assumed pollutant type	Ratio of CAC cost to least cost
Atkinson and Lewis (1974)	Particulates	St. Louis Metro. Area	SIP regulations	Nonuniformly mixed assimilative	6.00[a]
Roach, et al. (1981)	Sulfur dioxide	Four Corners in Utah, Colorado, Arizona, and New Mexico	SIP regulations	Nonuniformly mixed assimilative	4.25
Hahn and Noll (1982)	Sulfates	Los Angeles	California emission standards	Nonuniformly mixed assimilative	1.07
Krupnick (1983)	Nitrogen dioxide	Baltimore	Proposed RACT regulations	Nonuniformly mixed assimilative	5.96[b]
Seskin, Anderson, and Reid (1983)	Nitrogen dioxide	Chicago	Proposed RACT regulations	Nonuniformly mixed assimilative	14.4[b]
McGartland (1984)	Particulates	Baltimore	SIP regulations	Nonuniformly mixed assimilative	4.18
Spofford (1984)	Sulfur dioxide	Lower Delaware Valley	Uniform percentage reduction	Nonuniformly mixed assimilative	1.78
	Particulates	Lower Delaware Valley	Uniform percentage reduction	Nonuniformly mixed assimilative	22.0
Harrison (1983)	Airport noise	United States	Mandatory retrofit	Uniformly mixed assimilative	1.72[c]
Maloney and Yandle (1984)	Hydrocarbons	All domestic DuPont plants	Uniform percentage reduction	Uniformly mixed assimilative	4.15[d]

| Palmer, Mooz, Quinn, and Wolf (1980) | Chlorofluoro- carbon emis- sions from nonaerosol applications | United States | Proposed emission standards | Uniformly mixed accumulative | 1.96 |

Definitions: CAC = Command and control, the traditional regulatory approach.

SIP = state implementation plan.

RACT = Reasonably available control technologies, a set of standards imposed on existing sources in nonattainment areas.

Sources: Scott E. Atkinson and Donald H. Lewis, "A Cost-Effectiveness Analysis of Alternative Air Quality Control Strategies," *Journal of Environmental Economics and Management* vol. 1, no. 3 (November 1974) p. 247; Fred Roach, Charles Kolstad, Allen V. Kneese, Richard Tobin, and Michael Williams, "Alternative Air Quality Policy Options in the Four Corners Region," *Southwest Review* vol. 1, no. 2 (Summer 1981) table 3, pp. 44–45; Robert W. Hahn and Roger G. Noll, "Designing a Market for Tradeable Emission Permits," in Wesley A. Magat, ed., *Reform of Environmental Regulation* (Cambridge, Mass., Ballinger, 1982), tables 7-5 and 7-6, pp. 132–133; Alan J. Krupnick, "Costs of Alternative Policies for the Control of NO$_2$ in the Baltimore Region" (unpublished Resources for the Future working paper, 1983) table 4, p. 22; Eugene P. Seskin, Robert J. Anderson, Jr., and Robert O. Reid, "An Empirical Analysis of Economic Strategies for Controlling Air Pollution," *Journal of Environmental Economics and Management* vol. 10, no. 2 (June 1983) tables 1 and 2, pp. 117 and 120; Albert Mark McGartland, "Marketable Permit Systems for Air Pollution Control: An Empirical Study," (Ph.D. dissertation, University of Maryland, 1984) table 4.2, p. 67a; Walter O. Spofford, Jr., "Efficiency Properties of Alternative Source Control Policies for Meeting Ambient Air Quality Standards: An Empirical Application to the Lower Delaware Valley" (unpublished Resources for the Future discussion paper D-118, February 1984) table 13, p. 77; David Harrison, Jr., "Case Study 1: The Regulation of Aircraft Noise," in Thomas C. Schelling, ed., *Incentives for Environmental Protection* (Cambridge, Mass., MIT Press, 1983) tables 3.6 and 3.16, pp. 81 and 96; Michael T. Maloney and Bruce Yandle, "Estimation of the Cost of Air Pollution Control Regulation," *Journal of Environmental Economics and Management* (1984, forthcoming) table IV; Adele R. Palmer, William E. Mooz, Timothy H. Quinn, and Kathleen A. Wolf, *Economic Implications of Regulating Chlorofluorocarbon Emissions from Nonaerosol Applications*, Report #R-2524-EPA prepared for the U.S. Environmental Protection Agency by the Rand Corporation (June 1980) table 4.7, p. 225.

[a] Based on a 40 g/m^3 at worst receptor.

[b] Based on a short-term, 1-hour average of 250 g/m^3.

[c] Because it is a benefit–cost study instead of a cost-effectiveness study, the Harrison comparison of the command-and-control approach with the least-cost allocation involves different benefit levels. Specifically, the benefit levels associated with the least-cost allocation are only 82 percent of those associated with the command-and-control allocation. To produce cost estimates based on more comparable benefits, as a first approximation the least-cost allocation was divided by 0.82 and the resulting number was compared with the command-and-control cost.

[d] Based on 85 percent reduction of emissions from all sources.

increase in cost that results from using the command-and-control approach rather than the lowest cost allocation. To ensure comparability for each study, both the command-and-control allocation and the least-cost allocation are defined in terms of the same set of ambient standards (but *not* necessarily the same emission loadings).

Of the eight studies dealing with nonuniformly mixed assimilative pollutants, six find very large potential savings in abatement costs. If we omit the Hahn and Noll (1982) study, the study registering the *smallest* cost savings (sulfur dioxide control in the Lower Delaware Valley) finds that the command-and-control approach results in abatement costs that are 78 percent larger than necessary to meet the standards. In the Chicago study, the command-and-control allocation is estimated to be over *fourteen* times more expensive, and in the Lower Delaware Valley particulate study it is *twenty-two* times more expensive.

The Seskin, Anderson, and Reid (1983) and Spofford (1984) estimates are sufficiently larger than the others to deserve special mention. The former study is concerned with meeting a stringent, short-term standard. The concentrations at the most polluted receptors are heavily influenced by the location of the emissions. Because the least-cost strategy takes emission location into account while the command-and-control strategy does not, it is able to meet the ambient standard at a significantly lower cost.

The Spofford study obtains such a high potential cost savings for particulate control because of the manner in which it treats area sources in the command-and-control allocation. Application of a uniform percentage reduction to all area sources, as well as to point sources, turns out to be a very expensive way to control pollution. Some 89 percent of the total regional control costs associated with the command-and-control allocation are accounted for by area sources, while they account for only about 12 percent of total emissions.[2] Obviously, this is an extremely costly way to control area sources.

The relative cost savings for the uniformly mixed pollutants of both types, though modest by the standards of the other studies, are still large in an absolute sense. The command-and-control approach is estimated to be 72 percent more expensive than necessary in controlling airport noise, 96 percent more expensive in controlling nonaerosol applications of chlorofluorocarbons, and 315 percent more expensive in controlling hydrocarbons.

Similar studies have been done for nonuniformly mixed assimilative water pollutants. Because the control problem for these pollutants is so similar to that for air pollution, we can use these studies to broaden our qualitative knowledge of the determinants of the magnitude of potential cost savings.

2. See Spofford (1984, pp. 30 and 51).

The results of these studies are presented in table 5.[3] The cost savings are clearly there, but the magnitude for the average pollutant and geographic area is smaller for water than for air pollution. The largest ratio for water pollution is 3.13, compared with 22.0 for air pollution. Indeed, if the Los Angeles sulfate and Lower Delaware Valley sulfur dioxide studies are excluded, the *smallest* percentage cost savings for air pollution exceeds the *largest* percentage cost savings for water pollution.

The Causes and Consequences of Low Potential Cost Savings

Under what conditions might the potential cost savings be small? One clue can be found in the Los Angeles study, which estimated that the potential cost savings from using emissions trading in that city would be very small. A closer look at that study begins to suggest a circumstance in which the normally powerful cost-reducing properties of emissions trading may be less effective.

In Los Angeles, where a great deal of sulfate pollution is occurring, there were several reasons why the estimated cost savings were so low.[4] First, in contrast to other areas, the command-and-control strategy in California did not include scrubbers, a very expensive approach. Had California required scrubbers, the potential cost savings would have been higher.

Another reason, of more general applicability, is that the amount of control required is so great that every source is forced to control as much as is economically feasible. By definition, little further control can be undertaken. Since further reductions are necessary for credit transfers in emission reduction to take place, the opportunities for cost-saving transfers are extremely limited. Therefore, as long as the control authority has to impose emissions standards which are close to the limit of technological feasibility, in the absence of any changes in circumstances the divergence between the command-and-control and least-cost allocation would be small.[5]

3. One study apparently meeting the criteria for inclusion was excluded because its simulation of the command-and-control equilibrium *assumed* a considerable amount of cost minimizing in selecting the emission standards. See Jacobs and Casler (1979, p. 310).

4. Sulfates are not on the federal list of pollutants requiring state action. The decision to regulate sulfates and the choice of standards were the result of a unilateral California initiative.

5. Hahn and Noll (1982) also offer two further interpretations of their results. First they list several reasons for believing their permit cost data are biased upward (pp. 125–126). Principally, their data were derived from regulatory proceedings and suffer from any information problems regulators face. Second, they argue that "the local pollution control agency has attempted to use cost effectiveness as a major criterion in promulgating rules" (p. 131). In California, the Air Resources Board has a multimillion dollar research budget which it spends in part to gain good cost function information. Furthermore, these cost functions are used to set cost-effective standards.

Table 5. Empirical Studies of Water Pollution Control

Study and year	Pollutants covered	Geographic area	CAC benchmark	DO target (mg/liter)	Ratio of CAC cost to least cost
Johnson (1967)	Biochemical oxygen demand	Delaware Estuary—86-mile reach	Equal proportional treatment	2.0 3.0 4.0	3.13 1.62 1.43
O'Neil (1980)	Biochemical oxygen demand	20-mile segment of Lower Fox River in Wisconsin	Equal proportional treatment	2.0 4.0 6.2 7.9	2.29 1.71 1.45 1.38
Eheart, Brill, and Lyon (1983)	Biochemical oxygen demand	Willamette River in Oregon	Equal proportional treatment	4.8 7.5	1.12 1.19
		Delaware Estuary in Penn., Delaware, and New Jersey	Equal proportional treatment	3.0 3.6	3.00 2.92
		Upper Hudson River in New York	Equal proportional treatment	5.1 5.9	1.54 1.62
		Mohawk River in New York	Equal proportional treatment	6.8	1.22

Definitions: CAC = Command and control, the traditional regulatory approach.
DO = Dissolved oxygen: Higher DO targets indicate higher water quality.

Sources: Edwin L. Johnson, "A Study in the Economics of Water Quality Management," *Water Resources Research* vol. 3, no. 1 (Second Quarter, 1967) table 1, p. 297; William B. O'Neil, "Pollution Permits and Markets for Water Quality" (Ph.D. dissertation, University of Wisconsin-Madison, 1980) table 3.4, p. 65; J. Wayland Eheart, E. Downey Brill, Jr., and Randolph M. Lyon, "Transferable Discharge Permits for Control of BOD: An Overview," in Erhard F. Joeres and Martin H. David, eds., *Buying a Better Environment: Cost-Effective Regulation Through Permit Trading*, (Madison, Wis., University of Wisconsin Press, 1983) table 1, p. 177.

While the divergence has to be zero at the maximum control point (assuming the command-and-control allocation does not err by imposing infeasible reductions), it is not clear that the potential cost savings would decline monotonically with the degree of control required. Atkinson and Lewis (1974, p. 245), for example, show that for particulate control in St. Louis there is a considerable range of increasingly stringent ambient standards over which the potential cost savings actually increase.

Other studies generally have presented less complete information on the full range of possible ambient standards, concentrating instead on the more stringent end of the range. They generally find that in this range increasing stringency implies reduced potential cost savings. Spofford (1984, pp. 57, 66, and 77) finds this to be the case for both particulates and sulfur dioxide, while Maloney and Yandle (1984, table V) find it to be true for hydrocarbon control.

Of the five water pollution simulations in table 5 reporting more than one dissolved oxygen (DO) target, three find that the more stringent the standard (indicated by higher dissolved oxygen targets), the smaller the relative potential cost savings. The two studies finding the relative potential cost savings increasing with the stringency of the DO target estimate the increase to be very small.

THE POTENTIAL FOR ACTUAL COST SAVINGS

These results suggest that normally the command-and-control policy allocates the control responsibility in a manner that causes the control costs to be substantially higher than necessary. The potential for cost reduction is substantial.

It would be a mistake, however, to use these estimates of potential cost savings as if they were measuring the expected savings from the emissions trading program established by EPA. There are two main reasons why the actual cost savings are likely to differ from the estimates of potential savings: (1) the estimates of potential cost savings are subject to some error, and (2) the emissions trading program in existence differs markedly from the idealized programs modeled in chapter 2.

The Measurement Error Problem

The estimates of potential cost savings from simulation models are enormously useful when the simulations are to assist in deciding *whether* to allow emissions trading and, if so, *what forms* the program should take. For this purpose, what is important is the order of magnitude of the relative cost savings and how sensitive this order of magnitude is to various design changes. For the most part, the magnitudes have been so large as to sustain the desirability of reform, even taking the errors into account.

Nonetheless, it would be a mistake to use the simulated estimates of potential cost savings as a precise benchmark for evaluating specific emissions trading programs, even if the programs designed were completely in accord with the dictates of cost effectiveness. Perhaps the most compelling reason for not treating these estimates as a maximum achievable savings for an emissions trading program relates to the starting point for the analysis. The typical simulation comparing the command-and-control allocation with the least-cost allocation implicitly assumes that the starting point is no control at all. While this is an appropriate comparison in its own right, it is *not* appropriate to interpret the resulting cost difference as a measure of the potential cost savings that could be achieved by the EPA emissions trading program. The most this emissions trading program could save under the best of circumstances would be less than the potential cost saving because the reform package inevitably complements an *existing* command-and-control system rather than starting from scratch.

The distinction is important. When command-and-control regulations are in place prior to the time the emissions trading is introduced, sources would have already purchased and installed a good deal of durable capital equipment in order to comply with those regulations. Because this allocation of capital is fixed, it constrains the feasible set of reallocations of control responsibility that can be achieved by the reform package.

The simulation models do not assume a fixed, predetermined allocation of control equipment and, as a result, the least-cost solution may imply a quite different allocation of this capital. The computer estimates represent a correct calculation of the maximum cost saving from the reform package if and only if all the capital equipment in the command-and-control strategy could be, at no cost, disassembled at those sources needing less control in the least-cost solution and reassembled at those sources needing more. Since as a practical matter this cannot be done, the existing configuration of control equipment is a constraint on emissions trading which the models fail to recognize. Because emissions trading does not face a clean slate in assigning control responsibility, the maximum cost savings actually achieved would be smaller than predicted by the models.

The degree of overstatement would depend on how much capital equipment for control is already in place when the reforms are initiated in a particular state. The more capital equipment already in place, the larger the bias. To some extent this may depend on the pollutant. Because particulates and sulfur dioxide have been regulated somewhat longer, the degree of overstatement may tend to be larger for them.

In the *long run*—a period of time sufficiently long that the menu of control possibilities can change and sources can react to the enhanced set of possibilities—the cost savings calculated by these static models are likely to underestimate the maximum actual savings. This point is particularly important in interpreting one of our previous findings. When meeting the ambient standards is sufficiently difficult that maximum feasible control is necessary for every source, the potential cost savings from moving away from the command-and-control approach are zero *at that point in time.*

This evidence does not necessarily support the stronger conclusion that an emissions trading program would be useful only in those areas requiring less than full control and for those areas only during a transition period until full control is reached. Though this is a possible interpretation of what the simulation models are telling us, it ignores those dynamic effects of emissions trading that are not captured by the static models. Static models provide a snapshot of the situation at a moment in time while dynamic models follow the evolution of the system over time. Because of their timelessness, static models are incapable of capturing the role of transferable permits in influencing the evolution of the economic system (the *stimulating* role) and in responding to that evolution (the *facilitating* role). Both are enduring roles for emissions trading after the initial flurry of trading activity has passed.

Russell (1981) has attempted to assess the importance of the facilitating role by simulating the effects on permit markets of regional economic growth, changing technology, and changing product mix. Focusing on the steel, paper, and petroleum-refining industries in the eleven-county Delaware River Estuary Region, his study estimated the changes in permit use for three water pollutants (BOD, total suspended solids, and ammonia) that would have resulted if an emissions trading system were in place over the 1940–78 period. The calculations assume that the plants existing in 1940 would have been allocated permits to legitimize their emissions at that time, that new sources would have had to purchase emission reduction credit offsets according to current new-source rules, and that plant shutdowns or contractions would free up emission reduction credits for others to purchase.

This study found that for almost every decade and pollutant a substantial number of emission reduction credits would have been made available by plant closings, capacity contractions, product-mix changes, and/or the availability of new technologies. None of these opportunities for cost saving would have been picked up by a static model. Yet they are apparently important in the total life of an emissions trading program.

In the absence of an emissions trading approach, a control authority would have had not only to keep abreast of all technological developments so emission standards could be adjusted accordingly, but it would also have had to assure an overall balance between emission increases and decreases so as to preserve air quality. This tough assignment is handled completely by the market in an emissions trading program, facilitating the evolution of the system.

Emissions trading encourages as well as facilitates this evolution. Since emission reduction credits have value, in order to minimize costs sources must be continually looking for new opportunities to control emissions at lower cost. This search eventually results in the adoption of new technologies and the initiation of changes in product mix.

Emissions trading also encourages the earlier shutdown of plants which, for whatever reason, are likely to close sooner or later anyway. Plants that are shutting down can sell their emission reduction credits, producing an extra source of revenue that is eliminated under a command-and-control system. An interesting property of this revenue is that it would be higher in markets where the demand for emission reduction credits by expanded or new sources is higher. Therefore emissions trading would provide the most encouragement for shutdowns in those areas where other facilities are waiting to take up the slack. If the labor intensity of the new and old sources is the same, the impact of the shutdown on employment would be small because it would be offset by the gains in employment associated with the new source. Thus emissions trading would provide the most shutdown encouragement when the detrimental effects on employment were small.

In summary, emissions trading has a more permanent usefulness than static models suggest. Its usefulness is derived from its ability to facilitate, as well as to encourage, the evolution of the economy. The fact that emission reduction credits are valuable forces potential polluters to recognize that damage to the environment is a cost that must be paid and as such must become a consideration in household and business decision-making.

The Perfect Markets Assumption

The preceding section suggests two reasons why it would be a mistake to use the estimates of potential cost savings from the simulation models as forecasts of what will be achieved by the U.S. emissions trading program. On the one hand, the models assume that the least-cost allocation is unconstrained by any previous control decisions, whereas the emissions trading program was begun only after a great deal of control equipment was already in place. These past investments reduce the inherent flexibility in an emissions trading system and correspondingly reduce the savings for the economic life of that durable capital. On the other hand, because these models are static, reflecting only the current technological state of the art, they underestimate the potential cost savings in the long run as the new menu of control possibilities is developed.

Another reason why these estimates of potential cost savings cannot be used to forecast the magnitude of cost savings to be expected from the existing emissions trading program follows from the fact that the current program is a far cry from the idealized program being modeled. The models assume that there are regular markets for emission reduction credits, with stable prices and frequent transactions. In practice, transactions are infrequent and negotiated on a case-by-case basis. Despite the fact that the transactions costs associated with completing credit transfers for emission reductions are not negligible, they are ignored by the simulation model estimates. The models assume frictionless markets, whereas in fact a considerable amount of friction exists.

One source of friction arises in the implementation of these programs. To build constituencies as well as to protect against problems only dimly perceived in the abstract thinking about emissions trading, actual emissions trading is much more heavily controlled than the emissions trading envisioned in chapter 2. Though each of these topics will be covered in some detail in future chapters, it is important to capture some of the flavor of these restrictions to show that actual cost savings are likely to be smaller than those forecast by the models over the next few years.

- Cost-effective emissions trading involves many trades among sources that may not be located particularly close to each other. The current system discourages and in some cases prohibits these trades.

- Cost-effective emissions trading assumes that emission reduction credits can be used to meet any emission reduction standard. The current approach prevents their use to meet LAER, BACT, or NSPS standards.

- Cost-effective emissions trading assumes that emission timing is considered in allocating the control responsibility. The current program requires constant control and largely ignores the timing issue.

- Cost-effective emissions trading assumes that no participants can exercise any market power whereas current markets frequently contain a few large participants.

- Cost-effective emissions trading presumes that all allocations are equally easy to enforce and that complete compliance is forthcoming. In practice, the emissions trading program can affect the effectiveness of enforcement and the degree of compliance.

Actual permit markets are imperfect for two reasons: (1) the theory which we use to define perfection fails to incorporate all of the necessary elements of an emissions trading program, and (2) the bureaucracy is, on occasion, forced to deviate from perfection in order to gain support for the program. It is important to keep these two sources of imperfection separate because they imply rather different things. The former implies that the estimated cost savings are unrealistic because they are naive. The latter implies either that potential cost savings are being intentionally passed up in a quest for other goals or unintentionally passed up due to ignorance. Bureaucratic deviations are a particularly fruitful area for close scrutiny since they offer the possibility for further cost savings.

Emissions Trading Transactions

Though there is little doubt that the potential cost savings estimated from simulation models overstate the actual cost savings which could realistically be expected over the next few years, transactions data on emissions trading collected by EPA leave no doubt that a substantial cost savings is being achieved by the current emissions trading program. These data reinforce the conclusion that the costs associated with the command-and-control allocation are so excessive that the potential for actual savings is significant even when emissions trading is heavily controlled.

The existence of any emissions trading transactions, assuming equal enforceability, implies that the command-and-control allocation was not cost effective. If, contrary to chapter 2, the state control authorities were able to define cost-effective emission standards, there would be no incentive for sources to create or to use emission reduction credits. Each source would minimize its costs by accepting the responsibility assigned by the control authority.

There have been a large number of transactions suggesting that excessively expensive emission standards are common. As of September 1983, some 33 bubbles had been approved by EPA, 8 more were proposed, 9 were under review in Washington, 15 were under review at EPA regional offices, 49 were under review by states and at least 29 more were being actively considered by companies. In addition to these SIP revision trades, 14 bubbles were approved, 29 were being reviewed, and at least

6 were being actively developed under generic trading rules in 4 different states. The savings from the EPA approved and proposed bubbles is estimated at well over $200 million. EPA calculated that the total estimated savings resulting from all bubbles approved, proposed, or under development by September 1983 was over $700 million.[6]

To convey a sense of what kinds of transactions are involved, table 6 lists the bubbles approved by EPA during 1981, the first year in which bubble trades were approved. This table shows that during 1981, among those sources who reported the magnitude of their savings, some $8.45 to $8.55 million was saved in annual operating costs while $31.5 to $35.5 million was saved in capital costs. These are substantial savings even allowing for some possible overstatement. In many cases emissions as well as costs were reduced by the trades.

Not all pollutants are equally involved in trades. Table 7 summarizes the bubble trades by pollutant. These differences are not difficult to explain. Since bubble transactions are primarily being used to meet RACT requirements in nonattainment areas, that is where the demand is greatest. Since more areas are designated nonattainment for ozone than any other pollutant, more areas have an interest in trading pollutants involved in the formation of ozone. Volatile organic compounds are one of the main contributors to ozone pollution. In addition, because these are uniformly mixed assimilative pollutants, traders are not inhibited by the need to worry about the precise location of the emissions involved.

The second most common source for nonattainment designation is total suspended particulates, which is also the second most common pollutant involved in emission trading. In contrast, only one area, Los Angeles, has a serious nonattainment problem for nitrogen dioxide and the chief sources of the pollution are automobiles, not stationary sources.[7]

Volatile organic compounds also have the distinct advantage that because they are uniformly mixed assimilative pollutants, emission location makes no difference. Since the source generally bears the cost of monitoring and modeling for trades involving nonuniformly mixed assimilative pollutants, this is a significant cost borne during trades of other pollutants that volatile organic compound trades can avoid.

From 1976 to 1983 there have been at least 2,000 offset transactions.[8] The closest study of offset transactions has been done by Vivian (1981) and Vivian and Hall (1979). Of approximately 150 offset cases closely scrutinized, 55 were found to have led to emission reduction through the

6. These data are from Elman (1983).
7. National Commission on Air Quality (1981, pp. 113–115).
8. Palmisano (1983, p. 12).

Table 6. Bubble Trades Approved by EPA During 1981

Source	Industry category	Source of emission credit	Additional emission reduction	Cost savings
Narragansett Electric Providence, R.I.	Electric utility	Fuel switch	−317 lb/hr SO_2	$3 Million/yr fuel savings
Adolph Coors Boulder, Col.	Paper packaging	Change in control	No change VOC	$2.5 Million capital cost & $2–300,000/yr operating cost
3M Bristol, Pa.	Tape and packaging	Process change	−1,079 tpy VOC	$3 Million capital cost & $1.2 million/yr operating cost
McDonnell Douglas St. Louis, Mo.	Aerospace	Process change	−135 tpy VOC	Not available
Green River Station Muhlenberg, Ky.	Electric utility	Change in control	No change SO_2	$1.3 Million/yr operating cost
Armco, Inc. Middletown, Ohio	Steel	Change in control	−3,350 tpy TSP	$10–14 Million capital cost $2.5 million/yr operating cost
Andre's Greenhouse Doylestown, Pa.	Greenhouse	Fuel switch	No change SO_2	$250,000/yr operating cost

Company/Location	Industry	Type of change	Emission change	Cost
ITT Rayonier Jessup, Ga.	Pulp mill	Change in control	No change TSP	Not available
Old Crow Woodford, Ky.	Distillery	Change in control	−0.025 lb/MBtu TSP	Not available
3M Guin, Ala.	Glass manufacturing	Fuel switch	No change TSP	Not available
Uniroyal Naugatuck, Conn.	Chemical manufacturing	Fuel switch	No change SO_2	Not available
General Motors Defiance, Ohio	Foundry	Change in control	−34.8 lb/hr TSP	$12 Million capital cost
Shenango Allegheny, Pa.	Steel	Change in control	−207 tpy TSP	$4 Million capital cost
Corning Glass Works Danville, Ky.	Glass manufacturing	Change in control	−8.6 lb/hr TSP	Not available
Fasson-Avery Int'l Lake County, Ohio	Paper coating	Process change	No change VOC	Not available
Owens-Corning Fiberglass Newark, Ohio	Fiberglass manufacturing	Change in control	−17.18 lb/hr TSP	Not available

Definitions: SO_2 = Sulfur dioxide
VOC = Volatile organic compounds
TSP = Total suspended particulates

Source: Barry Elman, *Status Report on Emissions Trading Activity* (Washington, D.C., U.S. Environmental Protection Agency, November 1983).

Table 7.   Bubble Transactions by Type of Pollutant

Bubble application status	Type of pollutant				
	Sulfur dioxide	Volatile organic compounds	Total suspended particulates	Nitrogen oxides	Total
Approved	10	10	13	0	33
Proposed	1	2	5	0	8
Under review	3	52	18	0	73
Total	14	64	36	0	114

Source: Barry Elman, Status Report on Emissions Trading Activity (Washington, D.C., U.S. Environmental Protection Agency, November 1983).

use of improved technology and 15 resulted in emission reductions through fuel switching. Many offsets were made available when plants closed down.

The vast majority of offset and bubble transactions in all states have involved sources under a common ownership. It has been estimated that only about 50 offset transactions have involved credits created by one owner being transferred for consideration to another owner.[9] Furthermore, only two SIP revision bubble transactions prior to 1984 involved trades between plants controlled by different owners.

Emissions banking is in its infancy. As of September 1983, only 19 states and local districts had or were developing emissions banking programs. Four additional states were considering development of a bank. The majority of states did not yet have a banking program. One knowledgeable source estimated that fewer than a handful of trades had been facilitated by banking.[10]

Though the amount of information available on state experiences with the netting program is very limited, to date it probably has had very little impact. Not only is it the newest of the programs, but it has been under a legal cloud, which has caused states to be very cautious in using it until the legal issues had been resolved.

SUMMARY

• The potential for actual cost savings by emissions trading programs depends on how expensive the command-and-control allocation is relative to the least-cost allocation and how closely the allocation resulting from the emissions trading program approximates the least-cost allocation.

9. Ibid.
10. Ibid.

• A substantial majority, though not all, of the large number of computer simulation studies comparing the costs of command-and-control regulation of nonuniformly mixed assimilative pollutants with the least-cost allocation of control responsibility found the command-and-control responsibility to be significantly more expensive than necessary. Though fewer in number, studies of uniformly mixed accumulative pollutants and uniformly mixed assimilative pollutants find similar results. A single study shows the command-and-control allocation to be 96 percent more expensive than the least-cost allocation in controlling non-aerosol applications of chlorofluorocarbons. Two other studies found the command-and-control approach to be 72 percent more expensive in controlling airport noise and 315 percent more expensive in controlling volatile organic compounds than the least-cost allocation.

• Though the sample is rather small, the estimated potential cost savings for water pollutants (as a percentage of command-and-control cost) derived from those simulation models studied are generally also large; the average potential percentage savings for water pollutants is smaller than that for air pollutants.

• Whenever the need for additional reduction is so severe that the control authority has no choice but to impose emission standards which are close to the limit of technological feasibility, the immediate potential cost savings are very small.

• The estimates of potential cost savings derived from computer simulation models are not identical to the costs which could be saved even by perfectly designed emissions trading programs. In the short run, these models typically overestimate the maximum savings by assuming that any allocation of control responsibility is feasible. Whenever emissions trading programs are overlaid on an existing command-and-control system (as the U.S. program was), the durable control equipment already in place constrains the trading possibilities. In the long run these models underestimate the potential cost savings because they fail to account for the ability of emissions trading to stimulate, as well as to facilitate, the development and adoption of new control technologies.

• The emissions trading program established by EPA differs in a number of ways from the perfect markets modeled in the computer simulations. Characteristics of the current program which deviate from a cost-effective design include: (1) discouraging emissions trades between nonproximate sources, (2) prohibiting the use of emission reduction credits to meet the LAER, BACT, or NSPS standards, (3) ignoring the effect of emission timing on the achievement of the ambient standards, (4) creating the potential for market power, and (5) opening the door to enforcement and compliance problems.

• The existence of emissions trading activity is a sign that there are costs to be saved. Bubble and offset trading have been vigorous, involving a number of pollutants and a number of geographic areas. The estimated savings from all bubbles approved and under development is over $700 million.

• The pollutants which are most commonly involved in bubble trades are volatile organic compounds, followed by total suspended particulates and sulfur dioxides. One reason for this ranking is that these pollutants are the same three that are most commonly involved when the ambient standards are violated.

• Most bubble and offset transactions have involved transfers among emission points under common ownership. Transfers of credits among firms controlled by different owners account for only 2 or 3 percent of all offset transfers and less than 10 percent of all bubble transfers.

• Emissions banking is in its infancy. As of September 1983, only 19 states and local districts had or were developing emissions banking programs. This portion of the program has not yet had much of an impact in facilitating transactions.

• This evidence suggests that as currently configured, the emissions trading program can achieve air quality goals at substantially lower cost than the command-and-control policy. It also suggests that the savings could be even higher with different configurations. The next four chapters investigate specific implementation areas where this may be possible.

REFERENCES

Anderson, Robert J., Jr., Robert O. Reid, and Eugene P. Seskin. 1979. *An Analysis of Alternative Policies for Attaining and Maintaining a Short-Term NO₂ Standard* (Princeton, N.J., MATHTECH, Inc.).

Atkinson, Scott E., and Donald H. Lewis. 1974. "A Cost-Effectiveness Analysis of Alternative Air Quality Control Strategies," *Journal of Environmental Economics and Management* vol. 1, no. 3 (November) pp. 237–250.

Eheart, J. Wayland, E. Downey Brill, Jr., and Randolph M. Lyon. 1983. "Transferable Discharge Permits for Control of BOD: An Overview," in Erhard F. Joeres and Martin H. David, eds., *Buying a Better Environment: Cost-Effective Regulation Through Permit Trading* (Madison, Wis., University of Wisconsin Press) pp. 163–195.

Elman, Barry. 1983. *Status Report on Emission Trading Activity* (Washington, D.C., U.S. Environmental Protection Agency, November).

Hahn, Robert W., and Roger G. Noll. 1982. "Designing a Market for Tradeable Emission Permits," in Wesley A. Magat, ed., *Reform of Environmental Regulation* (Cambridge, Mass., Ballinger) pp. 119–146.

Harrison, David, Jr. 1983. "Case Study 1: The Regulation of Aircraft Noise," in Thomas C. Schelling, ed., *Incentives for Environmental Protection* (Cambridge, Mass., MIT Press) pp. 41–143.

Jacobs, James J., and George L. Casler. 1979. "Internalizing Externalities of Phosphorous Discharges from Crop Production to Surface Water: Effluent Taxes versus Uniform Reductions," *American Journal of Agricultural Economics* vol. 61, no. 2 (May) pp. 309–312.

Johnson, Edwin L. 1967. "A Study in the Economics of Water Quality Management," *Water Resources Research* vol. 3, no. 1 (Second Quarter) pp. 291–305.

Krupnick, Alan J. 1983. "Costs of Alternative Policies for the Control of NO_2 in the Baltimore Region" (unpublished Resources for the Future working paper).

Maloney, Michael T., and Bruce Yandle. 1984. "Estimation of the Cost of Air Pollution Control Regulation," *Journal of Environmental Economics and Management,* in press.

National Commission on Air Quality. 1981. *To Breathe Clean Air* (Washington, D.C., U.S. Government Printing Office).

Palmer, Adele R., William E. Mooz, Timothy H. Quinn, and Kathleen A. Wolf. 1980. *Economic Implications of Regulating Chlorofluorocarbon Emissions from Nonaerosol Applications.* Report #R-2524-EPA prepared for the U.S. Environmental Protection Agency by the Rand Corporation (June).

Palmisano, John. 1983. "An Evaluation of Emissions Trading." Paper presented at the 76th Annual Meeting of the Air Pollution Control Association, Atlanta, Georgia, June 23.

Roach, Fred, Charles Kolstad, Allen V. Kneese, Richard Tobin, and Michael Williams. 1981. "Alternative Air Quality Policy Options in the Four Corners Region," *Southwest Review* vol. 1, no. 2 (Summer) pp. 29–58.

Russell, Clifford S. 1981. "Controlled Trading of Pollution Permits," *Environmental Science and Technology* vol. 15, no. 1 (January) pp. 1–5.

Seskin, Eugene P., Robert J. Anderson, Jr., and Robert O. Reid. 1983. "An Empirical Analysis of Economic Strategies for Controlling Air Pollution," *Journal of Environmental Economics and Management* vol. 10, no. 2 (June) pp. 112–124.

Spofford, Walter O., Jr. 1984. "Efficiency Properties of Alternative Source Control Policies for Meeting Ambient Air Quality Standards: An Empirical Application to the Lower Delaware Valley" unpublished Resources for the Future discussion paper D-118 (February).

Vivian, W. 1981. "An Updated Tabulation of Offset Cases" cited in National Commission on Air Quality, *To Breathe Clean Air* (Washington, D.C., U.S. Government Printing Office).

Vivian, W., and W. Hall. 1979. "An Empirical Examination of U.S. Market Trading in Air Pollution Offsets," Institute of Public Policy Studies, University of Michigan.

4 / The Spatial Dimension

Sulfur dioxide, total suspended particulates and, to some extent, nitrogen dioxide are appropriately classified as nonuniformly mixed assimilative pollutants.[1] If the resulting allocation of control responsibility for these pollutants is to be cost effective, the theory reviewed in chapter 2 is convincing on the need for control authorities to consider *where* the pollutants are injected into the air as well as *how much*. Unfortunately, introducing source location into the policy design complicates matters; it is easier said than done.

DIFFICULTIES IN IMPLEMENTING
AN AMBIENT PERMIT SYSTEM

From a purely theoretical point of view, the challenge is easily manageable. All the control authority has to do is to implement the ambient permit system described in chapter 2. Unfortunately, the implementation of this system would not be a trivial matter. Both state control authorities and sources would have to overcome some rather formidable administrative and legislative barriers if an ambient permit system is to work smoothly in practice.

---

1. Nitrogen dioxide is controlled both as a direct contributor to health problems and as a contributor to ozone formation. With respect to the health effects, it is a nonuniformly mixed assimilative pollutant.

Transaction Complexity

The first such barrier is the inherent complexity of the ambient permit system. Because the law mandates that the ambient standards be met everywhere, complete assurance that violations would not occur requires a very large number of receptor locations. Fortunately, complete assurance is not required.

Reasonable assurance can be gained with relatively few receptor locations. Because any particular flow of pollutants will affect a number of sites, the readings at contiguous monitoring sites are highly correlated. Due to this interdependence, a small number of carefully placed monitors can give an adequate picture of pollutant concentrations over a fairly large geographic area. The studies investigating this question have typically found that nine or ten selected monitoring sites are adequate to cover a typical urban airshed.[2]

Though designing a transferable permit system to produce the desired concentrations at nine or ten sites is certainly more manageable than designing one for a much larger number of sites, it is a far from trivial exercise. In order to assure that trades do not jeopardize attainment at any of these monitoring locations, a separate market is required for each monitor. In particular, the traded credits would have to be defined in terms of the reduction in concentration achieved at each of the nine or ten monitor locations. Each of these monitor-specific, concentration reduction credits could be traded independently of the others. Fewer markets would leave some monitors unprotected, raising the possibility that trades would trigger violations at one or more of them.

The complexity of this system is illustrated by a description of what any particular source would have to go through to negotiate a trade. Suppose a source wished to expand its production facility in a non-attainment area. It could legitimize the resulting increase in emissions by purchasing sufficient concentration reduction credits from each affected monitor market. Because this proposed increase in emissions could be expected to affect concentrations at most, if not all, monitor locations, the source would be required to purchase a different number of offsetting credits in each of the markets. Each set of concentration reduction credits would command a different price, reflecting the difficulty of meeting the ambient standard at that monitor.

Since the emission increase is not legitimized until *all* required offsetting credits are obtained, the expansion could be jeopardized by problems in any one of these markets. Problems could arise, for example, if there were few sellers in one or more of the markets, since markets with few sellers provide less assurance that competitive prices will

2. See, for example, Ludwig, Javitz, and Valdes (1983).

prevail. When credit prices are not competitive, the transactions will not generally lead to a cost-effective allocation.[3]

Furthermore, in markets with few participants, prices may be more uncertain; prices for concentration reduction credits may be more commonly negotiated bilaterally on a case-by-case basis rather than determined by a market involving large numbers of buyers and sellers. When a source is required to negotiate in more than one market (as it is in an ambient permit approach), its problem becomes acute. The demand for credits in any one of the markets would depend not only on permit prices in that market, but on the prices in all other markets as well. The source could not definitively negotiate in market A until it knew the price in market B and vice versa. The interdependency among these markets creates an indeterminacy which can only be resolved in general by negotiating simultaneously in two or more markets at once. Though not impossible, this is a difficult burden for the source to bear.

This problem is exacerbated when control technologies are capable of controlling more than one pollutant.[4] In this circumstance, the desired number of credits of one type of pollutant will depend on the number of credits obtained for the other pollutant, and vice versa. Not only would the source be required to conduct simultaneous negotiations in different monitor markets of the *same* pollutant, it would also need to conduct simultaneous negotiations among the various markets associated with the different but related pollutants. The ability of sources to deal effectively with these interdependencies is questionable.

Control over Emissions

At several points in the Clean Air Act as well as in various legal cases interpreting it, a reduction in emissions is explicitly stipulated as the main means of achieving the ambient standards. Two examples are the requirement for "reasonable further progress" and the statutory provisions dealing with intermittent controls.

By creating nonattainment areas, Congress recognized and dealt with the (by then) obvious fact that these areas could not achieve compliance with the ambient standards for one or more pollutants by the previous statutory deadline. As one condition of extending these deadlines, in the interim Congress required "reasonable further progress" toward meeting the standards on an annual basis. For our purposes, the important characteristic of the requirement for "reasonable further progress" is

3. The problems associated with the existence of market power are explored more fully in chapter 6.

4. This kind of interdependency exists in controlling sulfur oxides and particulates. See Spofford (1984, pp. 95–97).

that it was defined in terms of emissions, not air quality.[5] Strategies which would improve air quality in nonattainment areas are prohibited by this provision if they simultaneously allow any increases in emissions. In nonattainment areas it is not permissible to rearrange the location of emissions to meet the standard, even when the standards could be met more rapidly and more cheaply that way.

Changing the location of emissions is not the only means of improving air quality without lowering emission rates that is ruled out by the Act. Dispersion techniques are also prohibited in all but extreme circumstances. These enhance the dilution of emissions so that by the time the pollution plume reaches the monitors, concentrations are relatively low. Two such techniques are tall stacks, which inject the pollutants into the air at a height high enough to disperse them by the time they hit the ground, and intermittent controls, which vary emission rates with atmospheric conditions (such as allowing higher rates on windy days). The Clean Air Act rules these approaches out to the extent they substitute for emission reduction.[6]

These statutory provisions place serious constraints on the ambient permit approach to pollution control. This approach capitalizes on source location and dispersion to improve air quality at a significantly lower cost than is possible when other approaches are used. Some trades under the ambient permit system would increase emissions. When sources that have a large impact on noncomplying receptors (those recording air quality levels that violate the standards) sell concentration reduction credits to sources having a large impact on receptors with significantly better-than-required air quality, emissions could increase. Despite the fact that these trades might reduce the cost of complying with the ambient standards as well as achieve more rapid compliance by concentrating the reductions on those monitors experiencing the violations, they would be ruled out by the current statutory requirements for emission reduction.

Although transaction complexity and the emission reduction requirements are both chinks in the armor of the ambient permit system, it is important to distinguish between them. Transaction complexity is an inherent difficulty with the ambient permit system, while the emission reduction requirements are constraints on the process resulting from

5. "The term 'reasonable further progress' means annual incremental reductions in emissions of the applicable air pollutant . . . which are sufficient in the judgement of the Administrator, to provide for attainment of the applicable national air quality standards by the date required . . . " 42 USC 7501.

6. See 42 USC 7423 for the statutory dispersion provisions and *Alabama Power Co.* v. *Costle,* 606 F. 2d 1068 (1979) for a court decision giving a strict interpretation of emission limitations. For a legal history and analysis of this provision, see Melnick (1983).

legislative action. Being neither inherent nor immutable, the emission reduction requirements could be changed by a legislative action that allows somewhat more flexibility.

The available empirical evidence suggests that this is not a trivial point. All studies of which I am aware that have examined this question have found that emission levels associated with the least-cost allocation are larger than command-and-control emission levels. The results of those studies are presented in table 8. Though the range is rather wide, it is clear that a substantial portion of the potential cost savings that could conceivably be achieved by an ambient permit system is directly attributable to the smaller emission reductions needed to meet the ambient standards when location and dispersion are taken into account. The current legislation rules out this portion of the savings.

While the ambient permit system may be a perfect theoretical solution to the problem of incorporating source location, in practice it is difficult to implement. Although the legal barriers could fall before a congressional modification of the Clean Air Act, the administrative complexity of the system is inherent.

POSSIBLE ALTERNATIVES

The vulnerability of the ambient permit system to these charges creates the need to examine alternative administratively and legally feasible approaches which, while they may not sustain the least-cost allocation, are less costly than the traditional approach. Four such approaches are considered here: (1) emission permit systems, (2) zonal permit systems, (3) single-market ambient permit systems, and (4) trading rules. One version of the last of these comes the closest to the current program, although the others have been considered by EPA at one point or another.

Emission Permits

One way to deal with the spatial complexity of pollution control is to ignore it. While from chapter 2 it is clear that an emission permit system would not support a cost-effective allocation of the control responsibility for nonuniformly mixed assimilative pollutants (only an ambient permit system could do that), the theory can provide no evidence on just how large the potential cost penalty would be. If it were small, emission permit systems might become an attractive approach. Because they are administratively simple and, by design, assure direct control over

Table 8. A Comparison of Emission Reduction for Command-and-Control and Least-Cost Approaches

Study and year	Pollutants considered	Geographic region	Command-and-control benchmark	Least-cost emission reduction as a percentage of command-and-control emission reduction[a] (percent)
Atkinson–Tietenberg (1982)[b]	Particulates	St. Louis	State implementation plan requirements	50.0
Seskin et al. (1983)	Nitrogen dioxide	Chicago	Proposed RACT standards	14.3
Krupnick (1983)	Nitrogen dioxide	Baltimore	Proposed RACT standards	51.6
McGartland (1983)	Particulates	Baltimore	State implementation plan requirements	92.2
Spofford (1984)[c]	Particulates	Lower Delaware Valley	Equal percentage reduction	83.0
	Sulfur dioxide		Equal percentage reduction	73.3

Sources: Scott E. Atkinson and T. H. Tietenberg, "The Empirical Properties of Two Classes of Designs for Transferable Discharge Permit Markets," Journal of Environmental Economics and Management vol. 9, no. 2 (June 1982) figure 5, p. 116; Eugene P. Seskin, Robert J. Anderson, Jr., and Robert O. Reid, "An Empirical Analysis of Economic Strategies for Controlling Air Pollution," Journal of Environmental Economics and Management vol. 10, no. 2 (June 1983) table 1, p. 117; Alan J. Krupnick, "Costs of Alternative Policies for the Control of NO_2 in the Baltimore Region," unpublished Resources for the Future working paper, 1983, table 4, p. 22; Albert Mark McGartland, "Marketable Permit Systems for Air Pollution Control: An Empirical Study," unpublished Ph.D. dissertation submitted to the University of Maryland, 1984, table 4.2, p. 67a; Walter O. Spofford, "Efficiency Properties of Alternative Source Control Policies for Meeting Ambient Air Quality Standards: An Empirical Application to the Lower Delaware Valley," unpublished Resources for the Future discussion paper D-118, table 21, p. 101.

[a] Each pair of allocations yields pollution levels which meet the same ambient standards.

[b] The 12 g/m³ standard is used for this calculation.

[c] Compares the least-cost strategy with the single-zone uniform percentage reduction.

emissions, they avoid the two previously discussed inherent weaknesses of the ambient permit system.

The evidence on the size of the potential cost penalty when emission permit systems are used to control nonuniformly mixed assimilative pollutants is presented in table 9. The potential abatement costs of an emission permit system are compared with those of the command-and-control and the ambient permit market (least-cost) allocations. In each study all three allocations of pollution control responsibility are defined such that they meet comparable ambient air quality standards.

In the fifth column of table 9 the potential abatement cost of the emission permit is compared with that of the traditional command-and-control approach. Because neither the emission permit allocation nor the command-and-control allocation are least-cost allocations for non-uniformly mixed assimilative pollutants, the cheaper allocation can only be identified empirically. A ratio of greater than 1.0 indicates that the emission permit approach achieves the objective at lower cost while a ratio of less than 1.0 indicates that the traditional regulatory approach is cheaper.

Perhaps the most obvious characteristic of these data is the variability of the results among various pollutants and regions. The ratios range from a low of 0.42 to a high of 11.10. The cost effectiveness of an emission permit system in this context is apparently quite sensitive to local conditions.

The difference in the cost of control resulting from the use of these two rather different approaches can be decomposed into two components: (1) the *equal-marginal-cost component* and (2) the *degree-of-required-control component*. The equal-marginal-cost component refers to the amount of the difference due to the equalization of marginal costs of control that would occur with an emission permit system, but not with the command-and-control approach. For any comparable degree of required reduction, the emission permit system would achieve that reduction at a lower cost. This component unambiguously favors the emission permit system.

The second component derives from the fact that the degree of required emission reduction is not usually the same for the two systems. Because the location of the sources matters, the degree of required emission reduction depends on the allocation of control responsibility among sources. Since the two systems result in different allocations of control responsibility among sources, the total amount of emission reduction needed to meet the ambient standards would not necessarily be the same.

The sign of this component is ambiguous. It can favor either the command-and-control system or the emission permit system, depending on which requires more control. How much control each requires can only be determined in a specific context.

In summary, whether the emission-permit or the command-and-control allocation is cheaper depends on the sign and magnitude of the degree-of-required-control component. If the command-and-control allocation requires more control, the emission permit system unambiguously results in lower control costs; both the equal-marginal-cost and degree-of-required-control components act in the same direction, reinforcing one another. Whenever the emission permit system requires more control, then the two components are of opposite sign and tend to offset each other. If the amount of reduction required in the emissions permit system is sufficiently large, the degree-of-required-control component would dominate the equal-marginal-cost component, causing the cost of control to be higher with an emission permit system.

Table 9 is of some help in identifying those local conditions which affect the interaction of these two components. One of those conditions is the stringency of the ambient standard relative to the level of uncontrolled emissions. Generally the higher the proportion of regional emissions that need to be controlled, the more the abatement costs of the two approaches converge. The convergence of these costs as the air quality standard becomes more stringent is clearly indicated in the Atkinson and Lewis (1974) study, for example.

A second clue is provided by the studies in which an emission permit system is actually more expensive than the traditional approach, a condition found in five of the eleven cases represented. (Though the dominance of the degree-of-required-control component was a clear theoretical possibility, it is rather striking how prevalent the phenomenon is, at least in these studies.) These studies are quite helpful in allowing us to discern the determinants of the degree of required control. The degree of emission reduction required by a permit system is quite sensitive to the spatial configuration of sources. When a few large sources are clustered near the receptor requiring the largest improvements in air quality, they would have to be controlled to a very high degree. Their resulting high marginal costs of control would be mirrored by equivalently high marginal costs of control for distant sources, despite the fact that emissions from distant sources have very little impact on the monitors where the greatest air quality improvement is needed. This overcontrol of distant sources results in much more emission reduction than necessary to meet the ambient standard under a command-and-control approach.

Other spatial configurations of sources require less overcontrol of distant sources in an emission permit system. When sources are more ubiquitous and no cluster dominates the most polluted receptor, a

Table 9. Using Emission Permit Systems to Control Nonuniformly Mixed Assimilative Pollutants: The Potential Cost

Study and year	Pollutants covered	Geographic area	CAC benchmark	Ratio of CAC to EPS abatement cost[a]	Ratio of EPS to APS abatement cost[a]
Atkinson and Lewis (1974)	Particulates	St. Louis metro. area	SIP regulations	6.00[b] 1.33[c]	1.67[b] 4.51[c]
Roach et al. (1981)	Sulfur dioxide	Four Corners in Utah, Colorado, Arizona, and New Mexico	SIP regulations	1.70	2.50
Hahn and Noll (1982)	Sulfates	Los Angeles	California regulations	1.05	1.07
Atkinson (1983)	Sulfur dioxide	Cuyahoga County, Ohio	SIP regulations	0.78[d] 0.91[e]	1.91[d] 1.40[e]
McGartland (1984)	Particulates	Baltimore	SIP regulations	2.50[f]	1.88
Krupnick (1983)	Nitrogen dioxide	Baltimore	Proposed RACT regulations	0.69[g]	8.64[g]
Seskin, Anderson, and Reid (1983)	Nitrogen dioxide	Chicago	Proposed RACT regulations	0.42	33.9

Spofford (1984)	Sulfur dioxide	Lower Delaware Valley	Equal percentage reduction	0.83[h]	2.13[h]
	Particulates			11.10[i]	1.97[i]

Definitions: CAC = Command and control, the traditional regulatory approach
EPS = Emission permit system
APS = Ambient permit system
SIP = State implementation plan
RACT = Reasonably available control technologies, a set of standards imposed on existing sources in nonattainment areas.

Sources: Scott E. Atkinson and Donald H. Lewis, "A Cost-Effectiveness Analysis of Alternative Air Quality Control Strategies," *Journal of Environmental Economics and Management* vol. 1, no. 3 (November 1974) p. 247; Fred Roach, Charles Kolstad, Allen V. Kneese, Richard Tobin, and Michael Williams, "Alternative Air Quality Policy Options in the Four Corners Region," *Southwest Review* vol. 1, no. 2 (Summer 1981) table 3, pp. 44–45; Robert W. Hahn and Roger G. Noll, "Designing a Market for Tradable Emissions Permits," in Wesley A. Magat, ed., *Reform of Environmental Regulation* (Cambridge, Mass., Ballinger, 1982) tables 7-5 and 7-6, pp. 132–133; Scott E. Atkinson, "Marketable Pollution Permits and Acid Rain Externalities," *Canadian Journal of Economics* vol. 16, no. 4 (November 1983) table 4, p. 716; Albert Mark McGartland, "Marketable Permit Systems for Air Pollution Control: An Empirical Study," unpublished Ph.D. dissertation submitted to the University of Maryland, 1984, tables 4.2 and 5.2a, pp. 67a and 77a; Alan J. Krupnick, "Costs of Alternative Policies for the Control of NO₂ in the Baltimore Region," unpublished Resources for the Future working paper, 1983, table 4, p. 22; Eugene P. Seskin, Robert J. Anderson, Jr., and Robert O. Reid, "An Empirical Analysis of Economic Strategies for Controlling Air Pollution," *Journal of Environmental Economics and Management* vol. 10, no. 2 (June 1983) tables 1 and 2, pp. 117 and 120; Walter O. Spofford, Jr., "Efficiency Properties of Alternative Source Control Policies for Meeting Ambient Air Quality Standards: An Empirical Application to the Lower Delaware Valley," unpublished Resources for the Future discussion paper D-118, tables 7 and 8, pp. 47 and 50.

[a] These columns assume emissions are reduced sufficiently by both policies to meet the ambient standards at all receptors. The ambient permit allocation is assumed to be identical to the least cost allocation.
[b] Assumes air quality of 60 g/m³ at worst receptor.
[c] Assumes air quality of 40 g/m³ at worst receptor.
[d] Assumes emission reduction sufficient to meet local ambient standards.
[e] Assumes emission reduction sufficient to meet local and long-range transport standards.
[f] Uses 100 g/m³ for EPS and 98 g/m³ for APS.
[g] Assumes air quality of 250 g/m³ at worst receptor.
[h] Assumes air quality of 80 g/m³ at worst receptor and both point and area sources controlled.
[i] Assumes air quality of 75 g/m³ at worst receptor.

permit system is able to achieve more balance between distant and proximate sources. In this circumstance, the air quality could be brought to the standard with both lower control costs and less total emission reduction.

It is possible, at least crudely, to test the hypothesis that the relative cost advantage (or disadvantage) of the emission permit system depends significantly on the amount of excess emission reduction that is required. If this hypothesis has merit, we should expect to find a significant cost disadvantage for the emission permit system in those studies where the amount of emission reduction required by the command-and-control approach is less than that required by the emission permit system.

Suppose that we were to rank the various studies in table 9 by the ratio of allowed permit emissions to allowed command-and-control emissions. If the hypothesis is valid, we should find a close positive correlation between this ranking and a ranking of the estimates based on the ratio of command-and-control to emission permit abatement costs. If the hypothesis is not valid, the correlation should be either zero or negative.

Table 10 presents the relevant ranks for all studies allowing the computation to be made.[7] The first noticeable aspect of these data is that of the five studies where the permit system abatement costs were higher than the command-and-control approach (ranks 1 through 5 in column 4), *all* require larger emission reductions than the command-and-control allocation. (Larger emission reductions are recorded as a number less than one in the first column.) In fact, larger emission reductions for the permit system are not uncommon; they occurred in eight out of the ten air pollution studies listed. The degree-of-required-control component apparently commonly favors the command-and-control approach.

It is possible to measure the correlation between these rankings using a rather well-known statistical device called Spearman's rank correlation coefficient. This coefficient can also be used to derive a statistic to test the significance of the measured correlation. If the hypothesis is correct, we should expect to reject the hypothesis that these rankings are either uncorrelated or negatively correlated. Since the Spearman rank coefficient is 0.87, the hypothesis that the true correlation is zero can be rejected with a 95 percent degree of confidence.

This result suggests that the amount of emission reduction required is a significant factor in explaining the cost effectiveness of the permit system. The fact that the correlation is not perfect, however, should not be overlooked, since this lack of perfection serves as a useful reminder that the test is a relatively crude one and that control costs are also important.

7. The Hahn and Noll (1982) study did not include a comparison of emission reductions.

Table 10.   Correlation of Ranks Between Relative Level of
Emission Reduction and Control Cost: Command-and-Control and
Emission Permit Systems

Study	Ratio of EPS emissions to CAC emissions	Rank	Ratio of CAC to EPS abatement cost[a]	Rank
Seskin, Anderson, and Reid (1983)	0.21	1	0.42	1
Atkinson (1983)	0.32	2	0.78	3
	0.39	3	0.91	5
Atkinson and Lewis (1974)	0.50	4	1.33	6
Krupnick (1983)	0.71	5.5	0.69	2
Roach et al. (1981)	0.71	5.5	1.70	7
Spofford (1984)	0.74	7	0.83	4
	0.84	8	11.10	10
McGartland (1984)	1.09	9	2.51	8
Atkinson and Lewis (1974)	1.11	10	6.00	9

Sources: See table 9.
[a] This column taken from table 9.

Notice in table 10 that two studies find that an emission permit system requires less emission reduction. As expected, they find that the control costs are lower for an emission permit system since the two components reinforce one another.

Also of interest are the five studies (ranks 6 through 10 in column 4) which show lower control costs for the emissions permit system. For three of these, the emissions permit system was cheaper despite requiring larger reductions. The equal-marginal-cost and degree-of-required-control components for these three studies tended to offset each other, but the dominance of the equal-marginal-cost component caused the emission permit system to result in a lower control cost.

Neither the command-and-control nor the emissions permit policy considers source location in assigning control responsibility. The magnitude of the cost penalty associated with ignoring source location in emission reduction credit trades can be ascertained by comparing the emission permit and ambient permit abatement costs, as is done in column 6 of table 9. It is easy to see that for every one of these studies, except the one by Hahn and Noll (1982), location matters. The cost savings lost by ignoring source location are large, even for those pollutants and regions where the emission permit system would be more expensive than the command-and-control allocation. By targeting the emission reduction on those sources having the largest impact on the binding receptors, less reduction in total emissions is required.

The increase in allowable emissions that normally accompanies policies incorporating source location is troublesome for pollutants that can be transported long distances. Ozone, sulfur oxides, and nitrogen oxides all fit in this category. For these pollutants, the computer-simulated cost savings may be misleading to the extent that they are based purely on the cost of meeting *local* receptors, allowing more emissions to be transported to other regions. Atkinson (1983) investigated the significance of this potential bias by comparing the cost saving attributable to incorporating location when only local receptors were considered to that when the contribution of emissions to long-range transport was also considered. The inclusion of long-range transport has two main effects: (1) it requires more total emission reduction and (2) it requires relatively more reduction from sources with tall stacks, since tall stacks enhance long-range transport. Atkinson's results indicate that although consideration of long-range transport tends to diminish the cost penalty associated with the emission permit system, it does not eliminate it. Even for long-range transport pollutants, the permit system still overcontrols emissions; location still matters though its influence is less significant than when only local receptors are considered.

Its normally large cost penalty is not the only strike against using an emission permit system to control nonuniformly mixed assimilative pollutants. Another arises from difficulties encountered in administering an emission permit system so as to fulfill the requirements of the Clean Air Act. Though this may seem odd, since the system appears to be very simple, this appearance is deceiving. It ignores the administrative problems encountered in initiating the system.

The Clean Air Act mandates that the ambient standards be met everywhere. By ignoring the location of the discharge point (the source of its simplicity), the emission permit policy forces the control authority to relinquish control over concentrations. Though it can control the total level of emissions, unfortunately there is no unique correspondence between the total level of emissions and air quality measured at the monitors. Concentration levels are sensitive to the location, as well as the amount, of emissions. Emission reduction credits (as opposed to concentration reduction credits) control only the latter.

The problem arises when the control authority attempts to issue the correct number of permits or, equivalently, to establish the control baseline used to define what reductions are surplus and therefore eligible to be certified as emission reduction credits. If it were perfectly omniscient, with full knowledge of the control costs of all emitters, defining the baseline would be a simple matter. With this detailed control cost information, it would be possible to anticipate how the credits would be traded in a market even before the market was initiated. Combining its presumed knowledge of control costs with its presumed knowledge of all transfer coefficients, the authority could define a control baseline which would just meet the ambient standard at the worst receptor. The calculated cost penalties in table 9 are based on just such an assumption about the behavior of this omniscient control authority.

But is that realistic? If the control authority were truly omniscient, there would be no need for permit markets to achieve cost effectiveness. It could mandate cost-effective emission standards for all sources directly without the bother of initiating permit markets. Indeed, it was the absence of this very information that triggered the interest in permit markets in the first place.

What is likely to happen in practice? Because the control authority would not normally know which sources would end up trading the emission reduction credits, it would, in all probability, build a "safety margin" into its calculation of the control baseline. By forcing more control than necessary to meet the ambient standards under conditions of perfect information, it could lower the likelihood of hot spots, areas with concentrations above the ambient standard. Due to this safety margin, the actual cost penalty associated with the emission permit system would be larger under realistic assumptions about control authority behavior than modeled in the simulation studies.

Even in those cases where the cost penalty may not be larger than estimated, the risk of violating the ambient standards at one or more locations is increased by trading activity. Suppose, for example, a particular allocation of pretrade control responsibility is consistent with the ambient standards. For that moment the control authority would have fulfilled its statutory obligations. However, as the number and composition of sources changed over time, emission reduction credits would be traded and this assignment would be rearranged. Any rearrangement involving increases in the number of credits held by those sources near binding receptors would jeopardize compliance. Nothing in the design of the emission permit system prevents these concentrations from exceeding the ambient standard.

A third strike against an emission permit approach stems from its inability to affect the location of new emission sources. Since prices for reduction credits do not vary with location in an emissions permit market, the cost of pollution control for any potential emission source does not depend on location within that market. Yet if that area is to meet the ambient standards, source location is frequently crucial. In a nutshell, the emissions permit affords too little protection to the ambient standards over the long run by sending the wrong signals to potential polluters making location decisions.

Not much empirical work has been done on the seriousness of this particular flaw, though one piece of EPA-funded research is directly relevant.[8] This research supports the conclusion that although environmental factors are growing in importance, they still rank somewhat lower than such traditional factors as access to markets and production, transport costs, and the characteristics of the site. Nonetheless, for heavy emitters environmental control costs can be quite important.[9] This evidence suggests that while environmental control costs may not be very important to many industries, they are important to those industries whose location has the largest potential impact on achieving the ambient air quality standards. Location incentives cannot be ignored even though they may not affect many industries very much.

Zonal Permit Systems

Another possible approach considered by EPA from time to time is a zoned emission permit system. In this approach, the control region is divided up into a specific number of zones, with each zone allocated a baseline control responsibility. While emission reduction credits can be traded within each zone on a one-for-one basis, trading among zones is prohibited.

This system has a certain surface appeal because it appears to respond to some of the problems which plague both the emission and ambient permit systems. Whereas the emission permit system overcontrols distant sources, the zoned permit system creates separate markets for distant and proximate sources. Whereas the emission permit system is vulnerable to the creation of hot spots, the zoned permit system attempts to lower this vulnerability by reducing the number of trades beween nonproximate sources. Whereas the ambient permit system requires each source to purchase permits from many markets, the zoned permit approach allows sources to operate in only one market.

In a crude way the creation of zones takes location into account. The necessity for ambient modeling is eliminated by restricting trades to sources within the same proximate area. As long as all sources within each zone are closely clustered, and stack heights are ignored (two very strong assumptions), all sources within each zone might be expected to have similar transfer coefficients. As long as the trading sources have

8. Stafford (1983).
9. For some of the sources included in the sample; for example, nonattainment areas were simply excluded from the list of possible location sites.

similar transfer coefficients, emission trades would not cause large changes in concentration at the relevant receptors.

Unfortunately this historic rationale is flawed on a number of grounds, which can be identified even before we turn to the empirical evidence. The inability of sources to trade permits across zonal boundaries restricts trading opportunities and reduces the potential for cost savings. This inverse relationship between cost savings and zone size sets up an inherent conflict in determining the appropriate zone size. To provide maximum protection against hot spots, the zones should be relatively small. On the other hand, by restricting trading opportunities, small zones raise costs.

The implementation of a zonal permit system places a larger burden on the control authority than the implementation of an emission permit system. With the zonal permit system, the control authority not only has to define the correct total control baseline (a task it also faces with an emission permit system), it must decide how to allocate this total emission reduction among the zones as well. Because emission reduction credits cannot flow across zonal boundaries, these administratively determined, initial control assignments set definite, permanent limits on the trading possibilities. Determining the share of the total emission reduction to be assigned to each zone is a crucial responsibility of the control authority in evaluating the cost effectiveness of this approach.

In principle, at any point in time there is an allocation among zones that minimizes the cost for a given configuration of zones.[10] However, to define that allocation, the control authority would have to know the control cost functions of every source. Because such omniscience is an unrealistic expectation for any control authority, zonal allocations will in practice deviate from these full-information allocations.

Unfortunately the more realistic limited-information allocations would extract an additional cost penalty. Allocating too much control responsibility to one zone and too little to another would raise compliance costs above the least-cost solution, even if the control authority were able to decide the correct total emission reduction for the region as a whole.

Even if the control authority were able somehow to make the cost-minimizing assignment of control responsibility among zones for a

10. This would be a least-cost allocation among the set of possible zonal allocations for a given set of zonal boundaries; it would not in general produce the regional least-cost allocation because the zonal permit system causes the marginal costs of *emission* reduction to be equalized across sources within each zone, rather than the marginal costs of *concentration* reduction. The zonal permit allocation of control responsibility would coincide with the regional least-cost allocation in general only if each zone contained one and only one source and each such source received its cost-effective allocation.

particular point in time, the normal evolution of the local economy would require changes in this assignment over time. Even if it were possible to derive reasonable procedures for making periodic changes in zonal control assignments to respond to this evolution, it would be difficult for sources to respond in a cost-effective manner as long as the ground rules changed in unanticipated ways.

This discussion has suggested two sources of a cost penalty in the design of zonal permit systems: (1) the administrative allocation of control responsibility to zones (coupled with the inability to correct this allocation by means of interzonal trades) and (2) the use of emission (rather than concentration) reduction trades within zones, given an assignment among zones. It has also suggested the need to assess the extent of the hot spot problem when the baseline control level is not sufficiently stringent to protect against violations of the ambient standards under all possible trades. Simulation models can add further clarification by calculating the magnitudes of the cost penalties and determining the seriousness of the hot spot problem.

We begin with a discussion of full-information simulations, those which presume omniscient control authorities. Both the least-cost total emission reduction and the least-cost assignment of this reduction among zones, given the particular zonal configuration in that simulation, are assumed. Though unrealistic in their treatment of control authority behavior, these studies do tend to show the potential for zonal systems under the most congenial circumstances. These serve as a benchmark for our subsequent discussions of limited-information zonal permit systems.

In this full-information approach, as the number of zones is increased (by reducing the size of each zone) the cost effectiveness of the policy must increase. Smaller zones not only mean less *within-zone* cost penalty, but the *between-zone* cost penalty is eliminated by assumption as well. Because the correct amount of control responsibility is assumed to be allocated to each zone, there is no need to trade credits across zone boundaries to reduce costs.

Because each zone in this system represents an independent functioning market, the interesting empirical question is how sensitive the remaining cost penalty is to the size of the market. The first study (Roach et al., 1981) to attack this question examined the effects of applying an emission permit system on a regional, state, or airshed level while another study, by McGartland (1984), examined the effects of creating multiple zones within an airshed. Together these studies encompass a wide range of market sizes.

For every large region the Roach et al. (1981, p. 44) study found that rather large reductions in the cost penalty could be achieved by reducing the size of the zones. If one emissions trading system were used for the

Table 11.   The Effect of the Number of Zones on the Potential Cost
Effectiveness of a Full-Information Zonal Permit Policy:
Particulate Control in Baltimore, Maryland

Number of zones	Annual potential control cost (millions of 1980 dollars)	Percent of cost penalty remaining
One	66.82	100.0
Three	59.94	74.9
Six	48.42	32.7
Nine	42.97	12.7
Fifteen	42.03	9.3
Least cost	39.49	0.0

Source: Albert Mark McGartland, "Marketable Permits for Air Pollution Control: An Empirical
Study" (unpublished Ph.D. dissertation, University of Maryland, 1984) table 5.2a, p. 77a.

entire multistate Four Corners region, the control cost was estimated to
be three to four times higher than if separate markets were created for
each of the region's airsheds. The higher cost of the single large market
is caused by the overcontrol of distant sources. To ensure compliance
with the ambient standards in all locations, larger regional emission
reductions are required. With multiple zones, the reductions can be
selectively concentrated on those zones where they are most needed.
Targeting the reductions reduces the costs.

As shown in table 11, the McGartland study found further gains from
multiple zones even within an airshed. According to this study, it takes
at least three, and possibly as many as six, zones to cut the cost penalty
in half. This is a discouraging finding because the larger the number of
zones, the more restricted the trading opportunities among sources.
Though this is not an important restriction when the administrative
allocations among zones are optimal, in the more realistic setting of
limited information it is crucial.

This pessimistic interpretation is borne out by the studies that have
considered limited-information zonal systems—those that attempt to
replicate realistic control authority behavior.[11] Rather than use optimal
conditions from the model to determine both the total emission reduction and its allocation among zones, these studies base these determinations on rules of thumb frequently used by control authorities.

Contrary to the expectation that small zone sizes would afford better
control over concentrations, they found that the hot spot problem could
be severe even with very small zones.[12] Close inspection of the results
indicates that different stack heights among sources within the same

11. These include McGartland (1984), Spofford (1984), and Atkinson and Tietenberg
(1982).
12. Spofford (1984, p. 82) and Atkinson and Tietenberg (1982, p. 120).

zone are a major reason for this discrepancy. When within-zone stack heights vary considerably, even contiguous sources may have very different transfer coefficients. Within-zone emission trades among sources with quite different transfer coefficients could produce rather drastic changes in measured concentrations. Since the Clean Air Act does not allow hot spots, control authorities would have to increase the amount of required emission reduction within each vulnerable zone to allow a margin of safety. This defeats one of the central purposes of a zonal permit system—the prevention of overcontrol.

These studies also discovered that the total cost of a limited-information system would be quite sensitive to the initial allocation of permits among zones. Several rules of thumb which might be used by a control authority to make these zonal allocations were examined. They included (1) equal percentage reductions based upon uncontrolled emissions, (2) equal percentage reductions in currently allowed emissions, and (3) reductions based on the need to improve air quality at the nearest within-zone receptor. All these had large cost penalties associated with them; none emerged as particularly superior or desirable. All conventional administrative approaches to allocating permits among zones seem to undermine the usefulness of the zonal permit approach.

Unfortunately while these studies found that smaller zone sizes did not alleviate the hot spot problem very much, they also found that they increased the abatement costs significantly.[13] In limited-information permit systems, the conventional initial allocations produce high cost penalties which can be reduced only by trading. Smaller zones restrict trading opportunities substantially (since no trades are permitted across zone borders) and the costs rise accordingly.

In realistic circumstances, zonal permit systems do not appear promising. Because region-wide trades are an important source of cost reduction, the best systems must allow region-wide trades while not allowing hot spots to arise. The message from the simulation studies is that zonal permit systems do not provide much of an opportunity either to reduce costs or to control the hot spot problem.

Single-Market Ambient Permit Systems

One possible means of allowing region-wide trades while providing some protection against hot spots involves using a simple version of the ambient permit system in which all trades are consummated on the basis of their effect on a single "worst case" receptor. Other receptors are presumed to benefit indirectly from the emission reductions needed to meet the ambient standard at this receptor. The problem of transaction com-

13. Atkinson and Tietenberg (1982, p. 119) and Spofford (1984, p. 81).

plexity is avoided in this version since there is only one trading market for concentration reduction credits. Because region-wide trades are permitted, the cost penalty associated with restricted trading opportunities is avoided.

The degree of protection against hot spots afforded by this approach depends on local circumstances. Because the concentration reduction credits are defined in terms of a single receptor, that receptor is adequately protected, but the others are protected only indirectly. As long as the air quality in a particular region is dominated by a single receptor, in the short run this approach would typically impose only a small associated cost penalty while offering a substantial reduction in the complexity of compliance. The indirect protection would be adequate.

Adequate protection from hot spots all hinges on the dominance of a single receptor. Single-receptor dominance can be tested using the simulation models by examining the number of binding air quality constraints in the ambient permit solution. If single-receptor dominance exists, the air quality constraint will be binding at only one receptor. The air quality at the other receptors would be better than required. The price for the concentration reduction credit associated with the binding receptor would be positive, while the prices of the credits associated with the other receptors would be zero, reflecting their excess supply.

In general, the air simulation models find from one to three binding receptors. The single permit market would be fully cost effective in those cases with only one binding receptor; it would not in the others. Therefore to guarantee compliance at these other locations, more total emission reduction would be required. This additional control adds to the cost penalty. The only study of how large the cost penalty would be in cases with multiple binding receptors (Atkinson and Tietenberg, 1982, p. 114) concludes that it remains very small. In this study at least one receptor can serve as a useful proxy for the others, given the current configuration of sources.

This optimistic finding that a single market for concentration reduction credits would suffice is tempered to a considerable degree by two considerations. First, a single study is hardly conclusive. In the case of multiple binding receptors, the size of the amount of excess emission reduction needed to assure compliance would depend on the proximity of the binding receptors to each other, which would vary from study to study. Contiguous binding receptors would have lower cost penalties because one receptor can act as a reasonable proxy for the others. When the binding receptors are far apart, however, further reductions in concentrations at one will have only a minimal impact on the other. Therefore a policy geared toward one receptor could not be expected to deal effectively with the concentration levels at more distant receptors.

The second consideration is even more troublesome because it is an inherent flaw, and not one which depends simply on local conditions such as the number of binding receptors. A single market for concentration reduction credits would establish locational incentives which would eventually undermine its claim to reasonableness, even in those areas initially having only one binding receptor.

With a single market for concentration reduction credits, credits purchased for use at some distance from the receptor cost less per unit of emissions than credits for use near the receptor. For new sources, this creates an incentive to locate operations away from the initially binding receptor because credits needed to legitimize emissions would be cheaper there. When sufficient new sources have located near some new, remote receptor, that receptor would become binding as well. Because this trading system is based on preventing violations only at the initially binding receptor, it creates, not only the opportunity, but the incentive for other receptors to be violated as well. When this happens, the amount of required additional emission reduction, and, hence, the cost penalty, would rise since in all likelihood the new receptors would be distant from the receptor on which the single market is based.

Should a second receptor become binding, it is possible to initiate another market so that there would be two markets instead of one. Though two markets are still preferable to the nine or ten markets (one for each receptor) required by the full ambient permit system, the simplicity advantage diminishes over time. As more and more receptors become binding, more separate markets would have to be initiated. As time passes, the complexity of this version approaches that of the full ambient permit market. Since the appeal of this approach diminishes over time, implementing it now would ignore the inevitable complexity just around the corner. The desirability of the single-market ambient permit system is purely transitory.

Trading Rules

And so we finally come to what is probably the most practical way to incorporate source location into permit-based pollution control policies. In a sense the trading rules approach is an amalgam of many of the best aspects of the other policies. Because it avoids zones, region-wide trades are permitted. The hot spot problem is eliminated without recourse to multiple markets by the selective use of transfer coefficients.

The trading rule approach represents a departure from the previously discussed approaches because it is not based on a complete set of markets, each with predetermined prices. Trading rules specify the procedures for deciding how much of an increase in emissions is allowed a purchaser or combination of purchasers, given an emission reduction by a seller or set of sellers. They focus on the *transaction* on a case-by-case basis rather than focusing on the *market* as a whole. This approach presupposes that the attempt to define and market a separate concentration credit for each receptor is not likely to succeed. Yet it attempts to retain the use of transfer coefficients to govern trades among sources.

Three different trading rules have been proposed in the literature: (1) the pollution offset,[14] (2) the nondegradation offset,[15] and (3) the modified pollution offset.[16] The *pollution offset* approach allows offsetting trades among sources as long as they do not violate *any* ambient air quality standard. The *nondegradation offset* allows trades among sources as long as no ambient air quality standard is violated *and* total emissions do not increase. The *modified pollution offset* allows trades among sources as long as neither the pretrade air quality nor the ambient standard (whichever is more stringent) is exceeded at any receptor. Total emissions are not directly controlled.

Consider the effect of these trading rules on two representative sources contemplating a trade (figure 3). The trading possibilities are bounded by the various emission and air quality constraints. The pretrade situation is given as (E_1, E_2), the currently allowed emission rates for the two sources. R_2^S and R_1^S are, respectively, the emission combinations between the two sources which allow the ambient air quality standard to be met at the second and the first receptors. As drawn, in the pretrade situation only the second receptor is binding. R_1^C defines the allowable emission combinations if current air quality (which, by construction, is cleaner than required by the standards) is to be preserved at the first receptor. The 45° line records the allowable emission combinations which hold the level of total emissions constant at their pretrade level.

In figure 3 if the second source were to sell to the first source, the trading possibilities would be area A, if the modified pollution offset is used, $A + B$ if the nondegradation offset is used, and $A + B + C$ if the pollution offset rule is used.

Notice that if the second source is buying emission reductions, all three trading rules yield the same set of trading possibilities (area D). The equivalence of the trading possibilities for all three rules in this context derives from the fact that current air quality at the receptor most affected by the purchasing source is already at the standard. Emissions would decrease and air quality would remain at the standard regardless of which of the three rules governed the trade.

14. Krupnick, Oates, and Van de Verg (1983).
15. Atkinson and Tietenberg (1982).
16. McGartland and Oates (1983).

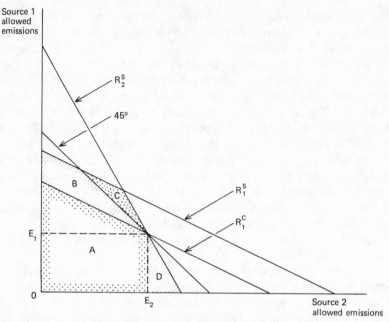

Figure 3. A comparison of three trading rules

Because costs are lowered as one moves away from the origin of this graph (since less control is required), after-trade allowable emissions would be represented by a point on the outer edge of the appropriate frontier. Those rules offering more trading possibilities would normally support lower cost allocations of the control responsibility. In this example, therefore, costs would be lowest for the pollution offset and highest for the modified pollution offset.

This potential cost advantage is one of two substantial advantages the pollution offset has. The second is that this is the only trading rule of the three which does not make the trading possibilities contingent on the pretrade emission situation for these two sources. This independence is clear from figure 3. R_1^S and R_2^S are a function of the ambient standards and the amount emitted by sources other than the two being considered; they are not affected by the command-and-control allocation in any way. However, both the 45° line and R_1^C are drawn *given a particular starting allocation between these two sources*. In the case of the U.S. emissions trading program, this starting allocation is the command-and-control allocation. R_1^C and the 45° line must pass through (E_1, E_2).

This point is significant because it means that the degree to which the nondegradation offset and the modified pollution offset can be expected to diverge from the least-cost allocation is sensitive to this initial, administratively determined allocation. In essence, when the command-and-control allocation defines the pretrade equilibrium, these trading rules perpetuate excessive control. Since one of the clearest conclusions that comes from chapter 3 is the large cost penalty associated with this excess emission reduction, its importance to the nondegradation and modified pollution offset trading rules is an unfortunate characteristic. Although the cost penalty associated with these two trading rules would still be less, usually substantially less, than the command-and-control pretrade allocation, it would never be eliminated except by coincidence.

Unfortunately the attractiveness of the pollution offset rule is marred by two other considerations: (1) any actual sequence of trades may fail to capture the potential cost savings and (2) for some trades the ratio of emission increases to emission reductions is not adequately defined by the rule.

Whereas Krupnick, Oates, and de Verg (1983, p. 24) show that in the two-source, two-receptor case all gains from trade will be exhausted so that the pollution offset trading equilibrium would be cost effective, it is not obvious that that result extends to the more realistic multiple trade context. When a source reduces emissions to create an emission reduction credit, it triggers concentration reductions at a number of receptors. Any source increasing emissions will (by acquiring credits) similarly trigger concentration increases at a number of receptors. Even if these *emission* increases and decreases were of the same magnitude, in general the *concentration* increases and decreases would not be of the same magnitude at every receptor.

In the pollution offset system, depending on the location and stack heights of the trading sources, an emissions trade would normally create a vector of net concentration increases and reductions at various receptors. By design, only concentration increases triggering a violation of the ambient standard would have any bearing on the trade. Once the trade had been consummated, the trading parties would lose any responsibility for or claim over these concentration changes.

Compare this situation with an ambient permit system in which the seller has a well-defined property right over all of the concentration reductions resulting from its emission reduction. Any selling source could treat concentration reduction credits as separate, tradeable commodities, retaining those not specifically traded to the acquiring source. The retained credits could be sold to other sources as needed.

Whereas the ambient permit system retains the value of these reductions to the creating source, thereby encouraging their transfer to those sources valuing them the most, the pollution offset system creates a "use it or lose it" situation. The property rights in the concentration reduction credits not needed in the specific trade at hand are lost. Only those concentration credits explicitly desired by the acquiring source would receive full value. Because the rest cannot be sold separately to other buyers, there is no guarantee that they would be transferred to those sources valuing them the most. Only if a series of transactions are consummated simultaneously and the valuable excess reductions are all acquired can this system result in a cost-effective allocation.

A second problem relates to defining the *offset ratio,* the amount of emission reduction any expanding source has to acquire from another somewhat distant source in order to compensate adequately for each unit of expected emission increase. According to the pollution offset system, an acquiring source need only secure enough emission reduction credits to prevent violations of the ambient standard. Suppose a source locates in a relatively clean portion of a nonattainment area where the most affected monitors are recording air quality levels that are quite a bit better than required by the standards. Suppose further that the increased emissions from this acquiring source would make the air quality worse at these monitors, but would not trigger any violations of the ambient standard. In this case, according to the pollution offset approach, the source would not have to gain *any* offsetting reduction from another source since the only constraint on the trade is the prohibition against ambient standard violations. Sources affecting nonbinding monitors would be allowed to increase their emissions without securing *any* compensating reductions.

In essence, these two problems are related. In the first case the problem results from inadequately defined property rights over created concentration reductions, preventing their transfer to those valuing them the most. In the second case the problem results from the fact that no one holds property rights over the potential concentration units represented by the air quality that is currently better than the standards, allowing them to be allocated on a first-come, first-served basis. In both cases because the prevailing property right structures create perverse incentives, the trading equilibrium would not normally be cost effective.

The two remaining types of trading rules attempt to provide some resolution of the second problem, but not the first. The nondegradation offset disallows any trades which increase emissions, while the modified pollution offset holds air quality at all specified monitors at least at current levels or at the ambient standard, whichever is more stringent. Both can be seen as crude ways of rationing the limited amount of "excess" air quality at nonbinding monitors by requiring some minimum amount of compensating offsets that every new or modified source must acquire.

Of these two means, the nondegradation offset is the simplest because the accounting system is so much more straightforward. It has only to keep track of emissions and possible ambient standard violations. The modified pollution offset prevents trades that allow any concentration increases at any monitor. Except when emission increases are exported via a tall stack, any normal bilateral trade between two sources would normally make air quality worse at those monitors nearest to the acquiring source. To avoid this prohibition, bilateral modified pollution offset trades must simultaneously be coupled with other reducing and increasing sources so that the pretrade balance among monitors is retained after the trades. That is clearly possible in the sense that such multiple source trades exist, but the amount of coordination required as well as the need for these reductions and increases to coincide in time would pose a significant barrier to trading.

Despite the fact that there are reasons to believe that the cost savings from the pollution offset and the modified pollution offset rules are not likely to coincide with the maximum possible cost savings for those systems, it is instructive to compare the cost effectiveness and emission loading characteristics on the assumption that they do coincide. McGartland and Oates (1983, p. 15) found that for particulate control in the Baltimore region, the modified pollution offset system could achieve the pollution target at less than half the cost of the command-and-control approach, but it was still 70 percent more expensive than the pollution offset approach. Interestingly, both systems resulted in more emissions than the command-and-control system. The excess emissions created by the modified pollution offset trades were transported by local winds to the ocean and therefore did not degrade the air quality at local receptors. McGartland and Oates did not examine the nondegradation offset.

Atkinson and Tietenberg (1984) have examined all three systems in the context of particulate control in St. Louis. Since the size of the cost deviation from the least-cost allocation depends on the pretrade allocation of control responsibility for both the modified pollution offset and the nondegradation rule, three different initial allocations were considered. The results indicate that when the primary ambient standard is the target, the nondegradation offset is only slightly more expensive than the ambient permit system. For two out of the three initial administrative allocations, the cost penalty associated with the use of the nondegradation offset was less than 10 percent. The modified pollution offset was not only more expensive, it resulted in more emissions. Substantial savings were possible for all three trading rules compared with the pretrade command-and-control allocations.

This paper also points out that as it becomes more and more difficult to meet the standards (as in nonattainment areas), the cost penalty becomes somewhat more sensitive to the rule used in allocating the initial administrative control responsibility. In particular, it was discovered that one rule which took location into account reduced the cost penalty substantially (compared with the other considered rules) by targeting the emission reduction at those areas where it was most needed.[17] By targeting the reductions so as to produce the maximum reduction in concentration at the binding receptors, less total emission reduction is needed to meet the ambient standards. Since the non-degradation offset approach constrains trades to ensure that emissions do not increase above pretrade levels, higher pretrade allowable emission levels mean lower posttrade costs.

THE CURRENT EMISSIONS TRADING PROGRAM

The current approach relies on trading rules rather than markets, a direction this analysis suggests represents a reasonable compromise between cost effectiveness and administrative practicality. Nonetheless, the analysis also points out a number of ways in which current practice could be improved.

The offset policy states two requirements that any offset trade must satisfy: (1) emission reductions from existing sources in the area of the proposed source must be at a level to ensure reasonable progress toward attainment and (2) the emission offsets will provide a positive net air quality benefit in the affected area.[18] In general, the first condition has been interpreted as requiring greater than one-for-one offset ratios.[19] The question of whether the net benefit test means more than simply ensuring progress toward meeting the standards has been left to the states.

The bubble policy has a similar two-pronged test. In general, trades in nonattainment areas should involve no total emission increases, but there are allowed exceptions, such as when the increase can be shown to be consistent with reasonable further progress (due to additional decreases elsewhere). In attainment areas, bubble trades that increase emissions are allowed, but they would consume some of the allowable

17. This rule defined each source's control responsibility (R_j) as

$$R_j = E \frac{\sum_i a_{ij} UE_j}{\sum_i \sum_j a_{ij} UE_j}$$

where E is the required regional emission reduction, a_{ij} is the transfer coefficient relating emissions from the jth source to air quality at the ith receptor and UE_j is the uncontrolled level of emissions from the jth source (assuming it exercised no control at all).

18. 40 C.F.R. part 51, appendix S, IV A.

19. 40 C.F.R. part 51, appendix S, IV E.

increment.[20] The air quality test in the bubble policy requires a demonstration of air quality equivalence before and after the trade. The definition of equivalence has also been left to the states.

Different states have chosen different ways to approach this net air quality test. Some states, such as Washington, intend to approach the issue on a case-by-case basis; no specific rule is spelled out in advance.[21] Maine defines the test solely in terms of violating the ambient standards.[22] Other states, such as New Jersey[23] and Rhode Island,[24] require that trades not produce a *significant impact* in the area of the source increasing emissions. Significant impact is defined as a pollutant-specific, quantitative ceiling on the amount of concentration increase permitted at any receptor.[25]

California control districts have taken a more restrictive approach. In San Francisco, for example, a concentric zone approach is used to dictate the size of the offset ratio.[26] The ratio is defined by the San Francisco authorities as a function of the distance between seller and purchaser, where the specific nature of the function depends on the pollutant.

For example, the offset ratio is fixed at 2 to 1 for nitrogen oxide trades consummated between two sources less than 30 but more than 15 miles apart or for sulfur oxide trades involving sources more than 10 but less than 5 miles apart. This ratio drops to 1.2 to 1 for nitrogen oxide trades involving sources less than 10 miles apart or sulfur oxide trades involving sources less than 5 miles apart.

In Los Angeles the offset ratio is defined on the basis of the formula:

$$\text{Offset ratio} = a + b \times d$$

where $a = 1.2$ for any external offset, d is the distance in kilometers between the trading sources, $b = 0.0$ at distances of less than 8 kilometers, and $b = 0.01$ at distances equal to or greater than 8 kilometers.

Both of these approaches require larger offset ratios from distant sources than proximate sources as a means of protecting the air quality near the expanding source. Trades among distant sources will improve air quality at the seller's location, but may degrade it at the purchaser's location. To provide some (crude) compensation for this degradation,

20. 47 F.R. 15076 at 15082 (7 April 1982).

21. State of Washington, Implementation of Regulation for Air Contaminant Sources, chapter 173-403 WAC.

22. Rhode Island Air Pollution Regulation no. 8, section 1.6.

23. New Jersey Administrative Code Title 7, chapter 27, subchapter 18, part 18.3.

24. Most states use the significant impact thresholds defined in 40 C.F.R. part 51, appendix S, III.A.

25. Maine Department of Environmental Protection, chapter 113, *Growth Offset Regulation*, p. 94.

26. The specifics of these approaches are described in Liroff (1980, pp. 31–34).

large offset ratios are required. This is an excessively costly approach. Not only does it automatically rule out any trades involving constant or increasing emissions, it makes trades among nonproximate sources *uniformly* more difficult, regardless of the circumstances. A selective approach would differentiate trades improving air quality at the most adversely affected receptors from those which do not. By making trades much more expensive than necessary to meet the objectives, this blanket treatment eliminates many possible attractive trades.

Compare these California approaches, for example, with the non-degradation offset approach. In the latter, the control authority would calculate (or would require the sources to calculate using standard procedures) the air quality impact of the proposed trade. If a 1 to 1 offset ratio would improve air quality at the receptors where violations are occurring without jeopardizing the ambient standards near the purchasing source, an emissions trade would be allowed. If a 1 to 1 offset ratio would trigger a violation *anywhere*, the ratio would be adjusted upward until complete posttrade compliance was assured at all receptors. In this case the offset ratio is calculated for the involved sources on the basis of their special circumstances, not a universal rule that must apply to all sources. Not all distant source trades deserve such harsh treatment, though some may.

Consider one circumstance where the current emissions trading approach could be particularly detrimental. Suppose a source in a heavily polluted area wanted to sell emission reduction credits to a suburban source and an offset ratio of 1.0 would be compatible with the interests of both sources. If consummated, the trade would hold emissions constant, would lower the cost of compliance, and would improve air quality in the most polluted areas. Yet while the nondegradation offset policy would allow the trade, virtually every state would currently prohibit it.

In the way it treats the spatial dimension, the current emissions trading program deviates from cost effectiveness in three main ways: (1) by creating a presumption toward emission trades, the current policy perpetuates the excess emission reduction inherent in the command-and-control system; (2) by failing to give favorable treatment to trades involving reductions at the most polluted receptors, the current program provides too little incentive to attain the ambient standards within the statutory deadlines; and (3) by making trades among distant sources much more expensive than necessary, the current program has reduced the number of trading opportunities and increased compliance costs.

All three of these problems can be traced back to a single source: the definition of reasonable further progress in terms of emissions improvement rather than air quality improvement. This definition has been responsible for an excessive fixation on emission reduction in nonattainment areas and can get in the way of improving air quality where the improvements are most needed.

If reasonable further progress were defined in terms of increases in air quality at those receptors where the standards are exceeded, one programmatic improvement could follow immediately. Although the presumption toward emission trades could be retained (along with prohibiting trades triggering a violation of the ambient standards), emission increases could be specifically allowed in nonattainment areas under the right conditions. Whenever the creation of the offsets reduces concentrations at the most polluted receptors and the acquiring source would not trigger a violation at any receptor at an offset ratio less than one, emission increases might be allowed. These increases would provide an inducement to create emission reduction credits at sources having the largest impact on the most polluted receptors. This favorable treatment would end once attainment was reached.

This new focus on air quality improvements would also encourage states to stop requiring offset ratios to be greater than one in nonattainment areas. This practice would be replaced by the more flexible approach described above, which tailors the offset ratio to the circumstance.

SUMMARY

• Incorporating source location into an emissions trading program is a difficult but manageable proposition.

• Though theoretically able to meet the challenge, in practice ambient permit markets are not likely to be used because of their inherent complexity. In addition, they are inconsistent with the statutory prohibitions against trades that allow emission increases since ambient permit markets allow more emissions than other market approaches considered.

• Because they overcontrol distant sources, unconstrained emissions permit trading systems typically result in much higher compliance costs than the least-cost solution. The cost penalty is particularly large when sources close to the most adversely affected receptors must control so much that they are on very steep portions of their marginal cost curves. This condition is relatively common in air pollution studies. Because they do not directly control concentrations, unconstrained trades of emission reduction credits run a high risk of creating hot spots, pollutant concentrations which exceed the ambient standards at one or more points within the airshed. To provide a margin of safety for the ambient standards, the amount of required emission reductions must be increased, forcing even more overcontrol and yet higher costs. In general, an unconstrained emissions permit market is neither a reliable means of achieving the air quality goals, nor very cost effective.

• Full-information zonal permit systems afford more protection for ambient standards than emission permit systems and reduce compliance costs. However, control authorities must be omniscient if this approach is to work in practice; the information burden is unrealistically large. Limited-information zonal permit systems, those which can be initiated with reasonable amounts of information, are much less attractive. Because zonal emission reduction credits cannot be traded across boundaries, the cost penalty would be very sensitive to the initial allocation of baseline control responsibility among zones. No conventional rule of thumb for allocating this required emission reduction among zones comes close to the cost-effective allocation. Furthermore, largely because of the influence of tall stacks, even small zones do not afford adequate protection against trades which trigger a violation of ambient standards.

• Single-market ambient permit systems focusing on a single "worst case" receptor can typically come very close to the least-cost allocation with a minimum of transaction complexity. They can also afford quite a high degree of control over the hot spot problem, given a stable configuration of sources and the absence of multiple binding receptors. Unfortunately they also provide incentives for changing the spatial configuration of sources, undermining the selection of one particular receptor as the worst case. Although this can be compensated for by adding new markets for every binding receptor, this destroys the simplicity of the system, its most attractive feature. Since this approach sows the seeds of its own eventual destruction, its attractiveness is purely transitory.

• The establishment of trading rules which govern individual transactions (rather than permit markets which encompass all transactions simultaneously) promises to reduce cost penalties significantly below levels associated with command-and-control levels, while affording adequate protection from the hot spots problem.

• Of the three trading rules examined, the pollution offset is the closest to the ambient permit system. Although in theory it has the lowest potential cost penalty, in practice any likely sequence of trades would produce a trading equilibrium which is not cost effective. Its other disadvantage is that if implemented it would allow emissions to increase substantially beyond their command-and-control levels.

• The nondegradation offset is the closest of the three to the current policy because it disallows trades that increase emissions. The limited available empirical evidence suggests that in the one area where this approach has been studied, the cost penalty associated with this approach is small. Its chief disadvantage is that the size of the cost penalty depends on how control authorities allocate the pretrade control responsibility; inefficient pretrade assignments requiring larger-than-necessary emission reductions reduce the possibilities for further trade by limiting severely the regional allowable emissions level.

• The modified pollution offset affords the most protection for current local air quality (although it does allow emission increases from tall stacks). It shares with the nondegradation offset the weakness that the magnitude of its cost penalty depends on the initial command-and-control allocation, while it shares with the pollution offset the weakness that unreasonably complicated trades would be required for the trading equilibrium to be cost effective, given the air quality levels.

• The current emissions trading program deviates from cost effectiveness in three main ways: (1) by creating a presumption toward emission trades based on an initial command-and-control assignment of the control responsibility, the current policy perpetuates an excessive degree of emission reduction in nonattainment areas; (2) by not giving favorable treatment to trades involving reductions at the most polluted receptors, there is too little incentive to attain the ambient standards within the statutory deadlines; and (3) by making trades among all sources (but especially nonproximate sources) excessively expensive, the current policy has reduced the set of trading opportunities and raised the cost of compliance.

• A statutory change defining reasonable further progress in *air quality* rather than *emission* terms, coupled with a slightly modified version of the constrained emissions trading system currently in existence, could stimulate quicker and less costly compliance with the ambient standards.

REFERENCES

Atkinson, Scott E. 1983. "Marketable Pollution Systems and Acid Rain Externalities," *Canadian Journal of Economics* vol. 16, no. 4 (November) pp. 704–722.

————, and Donald H. Lewis. 1974. "A Cost-Effectiveness Analysis of Alternative Air Quality Control Strategies," *Journal of Environmental Economics and Management* vol. 1, no. 3 (November) pp. 237–250.

————, and T. H. Tietenberg. 1982. "The Empirical Properties of Two Classes of Designs for Transferable Discharge Permit Markets," *Journal of Environmental Economics and Management* vol. 9, no. 2 (June) pp. 101–121.

———— and ————. 1984. "Approaches for Reaching Ambient Standards in Non-Attainment Areas: Financial Burden and Efficiency Considerations," *Land Economics* vol. 60, no. 2 (May) pp. 148–159.

Hahn, Robert W., and Roger G. Noll. 1982. "Designing a Market for Tradeable Emission Permits," in Wesley A. Magat, ed., *Reform of Environmental Regulation* (Cambridge, Mass., Ballinger) pp. 119–146.

Krupnick, Alan J., Wallace E. Oates, and Eric Van de Verg. 1983. "On Marketable Air-Pollution Permits: The Case for a System of Pollution Offsets," *Journal of Environmental Economics and Management* vol. 10, no. 3 (September) pp. 233–247.

Liroff, Richard A. 1980. *Air Pollution Offsets: Trading, Selling and Banking* (Washington, D.C., Conservation Foundation, 1980).

Ludwig, F. L., H. S. Javitz, and A. Valdes. 1983. "How Many Stations are Required to Estimate the Design Value and the Expected Number of Exceedances of the Ozone Standard in an Urban Area?" *Journal of the Air Pollution Control Association* vol. 33, no. 10 (October) pp. 963–967.

McGartland, Albert Mark. 1984. "Marketable Permit Systems for Air Pollution Control: An Empirical Study" (unpublished Ph.D. dissertation, University of Maryland).

————, and Wallace E. Oates. 1983. "Marketable Permits for the Prevention of Environmental Deterioration" (unpublished University of Maryland Working Paper no. 83-11).

Melnick, R. Shep. 1983. *Regulation and the Courts: The Case of the Clean Air Act* (Washington, D.C., Brookings Institution) pp. 113–154.

Roach, Fred, Charles Kolstad, Allen V. Kneese, Richard Tobin, and Michael Williams. 1981. Alternative Air Quality Policy Options in the Four Corners Region," *Southwestern Review* vol. 1, no. 2 (Summer) pp. 29–58.

Seskin, Eugene P., Robert J. Anderson, Jr., and Robert O. Reid. 1983. "An Empirical Analysis of Economic Strategies for Controlling Air Pollution," *Journal of Environmental Economics and Management* vol. 10, no. 2 (June) pp. 112–124.

Spofford, Walter O., Jr. 1984. "Efficiency Properties of Alternative Control Policies for Meeting Ambient Air Quality Standards: An Empirical Application to the Lower Delaware Valley," Resources for the Future Discussion Paper D-118 (February).

Stafford, Howard A. 1983. "The Effects of Environmental Regulations on Industrial Location: Summary," unpublished Working Paper of the Department of Geography, University of Cincinnati, Cincinnati, Ohio (June).

5 / Distributing the Financial Burden

Emissions trading approaches to pollution control involve two phases: (1) an initial allocation of control responsibility and (2) an organized market or series of trading rules which allow surplus emission or concentration reduction credits to be transferred from one source to another. Defining the baseline control responsibility, the focus of this chapter, is important not only because it has a major effect on the distribution of the financial burden associated with pollution control, but also because under the proper conditions it offers the control authority a great deal of flexibility in how it distributes that burden without jeopardizing the cost effectiveness of the program.

This distributional flexibility is a two-edged sword. On the one hand, it allows the control authority to pursue a just or fair distribution of the costs and benefits. On the other hand, the distribution of the financial burden can become a political struggle in and of itself, with a majority coalition channeling the lion's share of the benefits to itself and a disproportionate share of the costs to a reluctant minority.

This chapter defines and explains the possibilities for affecting the distribution of financial burden with a transferable permit system, and uses this framework to understand the current emissions trading program as well as to examine alternatives to it. Following a discussion of the existing estimates of the costs of the command-and-control approach, a survey of the empirical evidence is used to analyze the possible ways permit systems could be used to affect this cost burden. Finally, the emissions trading program is evaluated and further reforms are proposed.

THE COMMAND-AND-CONTROL FINANCIAL BURDEN

Has the financial burden associated with the traditional approach been distributed in accordance with standard norms of fairness? To answer that question, we must define what standard norms we have in mind. In economics there are two: (1) vertical equity and (2) horizontal equity. In this context both refer to the distribution of benefits and costs of pollution control among households.

Vertical equity, in its most general definition, states that unequals should be treated unequally. In economics this general principle is conventionally taken to mean that the poor should not bear a disproportionate share of the burden. According to this norm, regressive distributions of the burden should be avoided. A burden is *regressively distributed* if it consumes a larger proportion of the incomes of poorer than wealthier households.

Horizontal equity refers to the equal treatment of equals. Though the basis for defining equals can vary from circumstance to circumstance, for our purposes we shall pay particular attention to the treatment of households with similar income levels in different parts of the country. In this context, horizontal equity is achieved when households receiving equal incomes, but located in different geographic areas, receive the same level of net benefits (the excess of benefits over costs).

The traditional command-and-control approach has three characteristics that shape its impact on financial burden: (1) it follows the polluter-pay principle, (2) it involves higher than necessary control costs, and (3) it imposes the most stringent controls on new sources or sources that have undergone major modifications. The polluter-pay principle dictates that polluters (as opposed to taxpayers) should bear the initial incidence of pollution control. The second and third characteristics imply that the initial financial burden is large and not shared equally among sources.

It is not a trivial matter to trace out the ultimate incidence of this burden on households since it requires some knowledge of the degree to which industrial sources can pass on their costs. Depending on the circumstances, these costs could be shifted forward to consumers (as higher prices), backward to employees (as lower wages or employment), or simply remain with stockholders (as lower dividends or capital gains). Determining whether the command-and-control approach satisfies the vertical and horizontal equity criteria presumes some knowledge of the ultimate, not merely the initial, incidence.

The Role of the New Source Bias

As discussed in chapter 1, the traditional command-and-control regulatory approach focuses mainly on new sources. Existing sources are subject to control only when the new source emission reductions are not sufficient to reach attainment. As a result of this focus, the pollution control outlays have typically been larger for new sources than for old, diminishing their ability to compete. One predictable consequence of this new source bias in the regulations is a delay in the entrance of new firms and a reduction in their market share compared with a situation where the regulations affected old and new plants to the same degree.

Existing plants have benefitted from this new source bias. Since new firms are the higher cost producers (due to their higher pollution control outlays), their profits would be bid to zero. Competition would bid their prices down until they would just cover average costs. Because of their lower costs, however, existing firms facing the same prices could turn a profit. This profit would represent a form of Ricardian rent which would not be eliminated by competition.[1]

Estimates of the Employment and Price Impacts

Because of this new source bias in the regulations, the detrimental impact on employment in existing plants has been rather small. The "Economic Dislocation Early Warning System" was set up by EPA to monitor plant closings and associated job losses where pollution control was alleged to be a factor in the closing. The data collected by this monitoring system (U.S. Environmental Protection Agency, 1982) from January 1971 through September 1982 suggest that a total of 154 plants were closed, involving a total of 32,749 jobs.

Though small in proportion to the total number of jobs nationally, these losses have been concentrated in a few industries and a few regions. Twenty-three percent of the plant closings and 34 percent of the jobs lost were in the primary metals industries. An additional 14 percent of the closings and 20 percent of the jobs were lost in the chemicals and allied industries. About 62 percent of the plant closings and 66 percent of the associated job losses were concentrated in the Northeast and the Midwest. Over half of these were in the Midwest.

Accounting for job losses in existing firms is not the only way to measure the employment impact. Another involves calculating the potential jobs lost from the failure of new sources to begin operations as a consequence of the new source bias. Though hard data are not available to perform this calculation directly, in terms of reduced employment growth, presumably the high growth areas of the South and Southwest have been the hardest hit by these regulations. Their economic growth is fueled by the arrival of new sources or the expansion of old ones. Both

1. Koch and Leone (1979) provide an interesting example of how existing plants in the tissue industry were actually more profitable with water pollution regulation than without because the regulation served to diminish competition from new sources. Examples for air pollution control can be found in Maloney and McCormick (1982).

sources of growth have been inhibited by the relative cost disadvantage that new or expanded sources face under the traditional approach.

On balance, the combined employment effects have probably been rather small and, according to some observers, may even have been positive, with the employment gains in those industries producing pollution control equipment outstripping the losses in industries forced to buy this equipment.[2]

Have the effects on the prices paid by the consumers of the products produced by the regulated industries been similarly small? The evidence in chapter 3 on the excessive control costs associated with the traditional approach suggests that price increases might be a more serious problem. How has this cost burden affected the poor?

There is a virtual unanimity among those who have studied the cost distribution of the traditional approach to controlling stationary source air pollution that it is regressive.[3] Transmitted through higher prices, the higher costs associated with pollution control burden the poor disproportionately because poorer families spend a higher percentage of their income on the affected commodities than the rich.

The benefits of controlling stationary sources are reaped primarily by those living in the largest metropolitan areas, especially, but not exclusively, those in the Northeast, Middle Atlantic, and North Central regions.[4] Since the costs tend to be distributed more uniformly among regions than the benefits, the command-and-control policy tends to violate the horizontal equity criterion by favoring urban over rural areas and northern areas over the South.

In summary, the command-and-control policy created a distribution of the benefits and costs that violated both the horizontal and vertical equity criteria by favoring the rich over the poor, the North over the South, and urban over rural residents.

The Potential Role for Emissions Trading

Because they offer the potential for lower control costs, transferable permit approaches also offer the promise of reducing this burden on the poor. Lower industrial control costs mean smaller price increases, and smaller price increases mean lower relative burdens on lower income households. When less of the income in lower income households has to be devoted to pollution control, more is left for food, housing, education, and other goods or services.

2. See Portney (1981, pp. 41–42).
3. See, for example, Gianessi, Peskin, and Wolff (1979) and Dorfman and Snow (1975).
4. See Gianessi, Peskin, and Wolff (1979).

Montgomery (1972) has shown that as long as all sources are price takers and region-wide ambient trades are permitted, the ability of the permit market to attain the cost-effective allocation of control responsibility is not affected by the initial distribution of control responsibility.[5] As long as the firm stays in business, the initial allocation of control responsibility has no effect on the ultimate degree of control chosen by the firm. This property is of considerable interest because it allows the control authority great flexibility in deciding how the financial burden for meeting the stationary source control requirements is to be allocated. Permit approaches allow considerations other than cost effectiveness to enter the picture.

This capability of allowing cost sharing without jeopardizing cost effectiveness is particularly useful when the sources which can most afford controls are not those which can most cheaply control the pollution. Under the command-and-control approach, for example, some types of sources were allowed to escape regulation because they were marginal industries and could ill afford the controls if they were to stay in business.[6] The possibility of controlling those sources to a cost-effective degree was lost, not because they were more expensive to control, but rather because the additional expenditures could not be afforded.

An emissions trading system allows these two situations to be separated. The source that pays for the control need not be the source that installs it. A source that can control its emissions cheaply, but cannot afford even those expenditures, can sell emission reduction credits to the other sources, using the proceeds to finance the additional controls. In this way the financial burden is kept low and cost-effective controls are installed.

ASSIGNING THE INITIAL CONTROL RESPONSIBILITY

There are many possible ways of assigning the initial control responsibility and each has its own unique effect on source expenditures. For our purposes it is helpful to classify them into two categories: (1) those involving financial transfers to or from the government and (2) those involving only financial transfers among sources. The former category includes revenue auctions and subsidies while the latter includes grandfathering and zero-revenue auctions. The EPA emissions

5. This theorem does not hold either in a zonal emission permit system where region-wide trades are not permitted (see chapter 4) or when sources can exercise some control over permit price (see chapter 6).
6. A number of examples of this point are given in chapter 8.

trading program falls into the latter category as one version of a grand-fathering approach.

Revenue Auctions and Subsidies

Auctions and subsidies represent two rather different approaches to assigning the initial control responsibility. In a revenue auction, the control authority would put a fixed number of allowable emission permits up for bid.[7] The number of available permits for sale would be dictated by the amount of pollution that could be allowed while ensuring that the ambient standards were attained.

To the acquiring source, these permits would serve exactly the same purpose as an emissions reduction credit (for uniformly mixed assimilative pollutants) or a concentration reduction credit (for nonuniformly mixed assimilative pollutants); each permit or credit would allow one or more units of emissions or concentration. Individual sources would submit bids relating the number of permits desired for each of several possible prices. The control authority would then calculate the price at which the number of permits sought would equal the number available. The highest bidders would receive a number of permits corresponding to the number of winning bids. The payments would go to the control authority.

With a *subsidy* approach, the government would initially legitimize all pollution by giving sufficient permits to all sources (existing and new) to cover their uncontrolled emissions. To obtain reductions in the pollution level, the government would have to purchase permits from the sources and retire them. As new sources entered the area, more permits would have to be purchased by the government to keep pollution levels at the ambient standards in the face of an increasing number of emitters. Subsidy payments flow from the government to the sources of pollution, whereas auction payments flow from the sources to the government.

Given equal amounts of uncontrolled emissions, these two approaches have very different effects on the financial burden borne by a representative source (figure 4), but not on the amount controlled. With a per unit subsidy of P for each unit of emissions controlled, the source would receive a payment equal to $A + B + C$ if it controlled all its emissions. Such a choice, however, would not minimize its costs. For all units controlled beyond q^*, the marginal cost of controlling these additional units exceeds the per unit subsidy. Therefore to minimize its cost, the source would choose to control only q^* emissions, spending B on control

---

7. There are a large number of possible types of auctions. For our purposes, the distinctions are not very important and therefore we avoid unnecessary complexity by ignoring them.

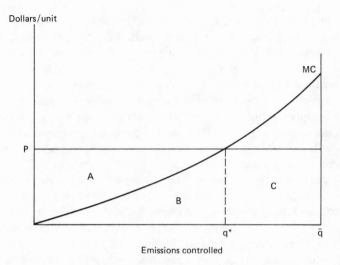

Figure 4. The size of the financial burden

equipment, but receiving $A + B$ from the government. Since the source receives more than it spends, A represents the profit.

In an auction market, the source would pay P for each purchased permit. A permit would be needed for each unit of emissions left uncontrolled. If the source chose not to control any emissions, it would spend all its money on permits. The resulting total cost, $A + B + C$, would be higher than necessary, however, since some emissions could be controlled for less money than spent on the permits. Therefore the source would purchase $\bar{q} - q^*$ permits, spending B on the control equipment and C on the permits.

Obviously expenditures by the representative source would be much greater in an auction market than when pollution control is subsidized. If the permit expenditures were large enough, they could even be larger than the control costs associated with the command-and-control regulation. With a subsidy, there is a sizable transfer from the government to the source, while in an auction market the transfer is from the source to the government. The command-and-control allocation involves no transfers.

Either of these approaches could eliminate the new source bias resulting from the command-and-control allocation by treating all sources (existing and new) alike. By subsidizing (or charging) new and old sources at the same rate, no competitive edge would be gained by existing sources.

Given our previous discussion of the regressive distribution of the cost burden for the traditional regulatory approach, it is clear that the subsidy approach would result in lower product prices. Lower product prices do not automatically imply less regressivity, however, unless the burden of the tax revenues used to finance the subsidy is progressively distributed. The tax burden is said to be *progressively distributed* when the tax payments comprise a higher proportion of larger than smaller incomes. The evidence seems to suggest that the tax burden is progressively distributed, and, therefore, for those areas of environmental policy where tax-financed subsidies are used, the regressivity is diminished.[8]

Subsidies also have serious drawbacks, however, which cause them to play a less central role than might have been expected based solely on their contribution to diminishing the regressivity of the pollution control cost burden. If the entire pollution control program were financed out of taxes, this would necessitate a sizable increase in the public budget. Never popular, even in the best of times, tax increases of the magnitude necessary to finance this budget increase seem particularly unlikely in this period of tax restraint.

Subsidies also produce rather unfortunate incentives. Not only do they encourage sources to choose processes with large amounts of uncontrolled emissions (making them eligible for large subsidies), but they send the wrong signals to sources about where to locate. For the subsidies to be cost effective, the highest per unit subsidies would have to be offered in those geographic areas where more stringent (higher marginal cost) control is needed. Sources considering locating in those areas would be attracted by the subsidies, since those areas would offer the highest profit per unit of control (area A in figure 4). The resulting immigration of new sources would make the control authority's job of securing the appropriate amount of pollution control tougher. Per-unit emission control subsidies make those areas where the costs of control are high appear the most desirable.

Grandfathering and Zero-Revenue Auctions

There are ways to initially distribute the control responsibility so as to diminish the regressivity of the control cost burden without taxing the public treasury and without creating perverse location incentives. They are based on the premise that any transfers to and from the government should cancel each other out.

Grandfathering is initiated by defining some baseline control responsibility—the degree of required emission reduction for each source. The

8. See Gianessi and Peskin (1980, p. 95).

source may fulfill its control responsibility either by reducing its own emissions at least as much as required (selling any additional reductions) or by acquiring enough credits from other sources to make up the difference between its actual and its required reduction. Because no transfers to or from the government are involved, on average, sources would pay only control costs; the sum of all permit expenditures would equal the sum of all credit receipts. For any individual source, actual outlays may exceed or be less than control costs, depending upon whether the source in question chooses to sell or to purchase credits.

One characteristic that differentiates auction and subsidy approaches from grandfathering is the manner in which the baseline control responsibility is specified. For auctions, the implied baseline control responsibility is for all sources to eliminate all emissions. Sufficient permits must be purchased from the government to cover the difference between complete and actual control. Subsidies imply the absence of any initial control responsibility; the government must pay for all reductions.

With grandfathering, the initial baseline control responsibility requires less than complete control, but enough to meet the ambient standards. Essentially the control authority decides how much reduction is needed and allocates this reduction among sources on the basis of some distribution rule.

The choice of a specific distribution rule can greatly affect the financial burden borne by individual sources, but it will have no effect on the financial burden borne by sources collectively. The chosen distribution rule would determine which sources would purchase permits and which would sell, but among all sources payments would equal receipts. The government has no financial stake in the outcome.

Zero-revenue auctions, originally proposed by Hahn and Noll (1982, p. 141), are a means of deriving the benefits of an auction without extracting large payments from sources. Essentially all collected revenues are rebated to the sources on the basis of a distribution rule which is, in principle, the same as that used in the grandfathering approach.

In the zero-revenue auction, each source would receive an initial distribution of credits based on this distribution rule, with the number of credits corresponding to the amount of pollution allowed by the ambient standards. All sources would be required to put the credits up for sale in an auction. Each source would then be required to place its bids for credits, specifying the number of credits desired at each of several possible prices. The control authority would collect these bids and determine the price at which the demand for credits would equal the supply. Sources would receive the number of credits corresponding to the number of credits they placed bids for at or over this market-clearing price. Their payment would equal $P(q_i^* - q_i^0)$ where P is the market-clearing price, q_i^* is the number of credits demanded by the ith source in the auction, and q_i^0 is its initial allocation of credits. Since, by construction,

$$\sum_i q_i^* = \sum_i q_i^0 ,$$

then

$$\sum_i P(q_i^* - q_i^0) = 0$$

Though any particular source may on balance pay more than it receives or vice versa, collectively the payments for the auctioned permits exactly equal the receipts in a zero-revenue auction.[9]

THE SIZE OF THE FINANCIAL BURDEN

The preference for grandfathering or zero-revenue auctions over a normal auction is based on both fairness and political feasibility considerations. The average industrial expenditures (counting permit or credit costs) are lower when no revenue is transferred to the government. This implies both lower price markups (to which the poor are seen as especially sensitive) and a presumed greater willingness on the part of sources to cooperate with a system that imposes lower costs on them.

The intensity of this preference depends on the magnitude of the financial burden imposed by permit or credit expenditures. Negligible financial burdens are not worth worrying about. Substantial financial burdens may eliminate some approaches from further consideration.

There are two measures of financial burden which are relevant: the total regional financial burden imposed on all sources and the distribution of this burden among individual sources. The former sheds light on the significance of permit or credit expenditures among the various approaches while the latter examines the effects of the various distribution rules on individual sources.

Regional Financial Burden

The regional financial burden is defined as the sum of control costs and permit or credit expenditures for all sources in the region. To show how

9. The astute reader at this point may wonder why the control authority should bother with a zero-revenue auction when it seems to have the same distributional consequences as a grandfathering system and yet is more complicated. The chief answer to this question relates to protecting against market power, a subject treated in the next chapter.

the various methods of initially allocating credits affect regional financial burden, it is convenient to express this definition symbolically as:

$$\text{regional financial burden} = \beta L + \alpha E$$
$$\beta \geq 1.0; \ -1 \leq \alpha \leq +1$$

where L is the minimum possible control cost for meeting the ambient standard, E is the maximum expenditure possible on credits, β is the ratio of the control cost for the approach being considered to the minimum control cost for the pollution standard, and α is a coefficient which expresses the degree of financial participation of the government.

Since appropriately defined emissions trading systems are cost effective ($\beta = 1$), their differences in regional financial burden are due to the value of α. For auction markets, $\alpha = 1.0$ because sources bear the entire permit expenditure burden collectively; every source has to purchase sufficient permits to legitimize its uncontrolled emissions. Subsidies involving the government purchase of permits from sources are represented by $\alpha = -1.0$. The grandfathering and zero-revenue auction options imply $\alpha = 0.0$ since for all sources taken together credit payments equal credit receipts.

Although from chapters 2 and 3 we know that an appropriately defined transferable permit system would yield lower control costs than the command-and-control allocation, total expenditures (including credit or permit expenditures) may or may not be higher. β is larger for command-and-control, but as long as α is positive, the payments for credits or permits may more than offset the control cost savings. Furthermore, for an inappropriately designed permit system (such as when an emission permit system is used to control nonuniformly mixed assimilative pollutants), it is not the case that control costs are minimized. Inappropriately designed emissions trading approaches can lead to larger financial burdens than the command-and-control approach both because they involve financial outlays on permits (which the command-and-control allocation does not), and because $\beta > 1$ for both approaches.

The last two columns of table 12 provide a direct comparison of the financial burdens of command-and-control and emissions trading for a variety of pollutants and regions when the credit expenditures are those which would result from an auction market ($\alpha = 1.0$). Consider first the ambient permit column. Though only three estimates are available, all indicate that total compliance costs are lower with this permit system than with a command-and-control allocation. For this type of permit market, if these studies are representative, the estimated savings in control cost commonly exceed the estimated permit expenditures.

Table 12. The Size of the Potential Regional Financial Burden:
Air Pollutants

| | | Ratio of permit expenditures to market abatement cost | | Ratio of total financial burden to CAC abatement cost | |
Study	Air or water quality level[a]	Ambient permit	Emission permit	Ambient permit	Emission permit
Nonuniformly mixed assimilative pollutants:					
Roach et al. (1981)					
Sulfur dioxide	1	n.a.	0.32	n.a.	0.59
Atkinson–Tietenberg (1982 and 1984)					
Particulates	1	0.52	3.00		
	2	0.72			
	3	0.85	5.67	0.42[b]	0.67[b]
	4	1.13	6.69		
	5	1.70	11.50		
	6	6.14	24.00		
Hahn and Noll (1982)					
Sulfates	1	n.a.	0.37	n.a.	1.09
	2	n.a.	0.41		
	3	n.a.	0.51		
	4	n.a.	0.32		
Seskin et al. (1983)					
Nitrogen oxides	1	0.44	1.36	0.10	6.08

Note: n.a. = Not available. N.A. = Not applicable.

Sources: Fred Roach, Charles Kolstad, Allen V. Kneese, Richard Tobin, and Michael Williams, "Alternative Air Quality Options in the Four Corners Region," *Southwest Review* vol. 1, no. 2 (Summer 1981) table 3, pp. 44–45; Scott E. Atkinson and T. H. Tietenberg, "The Empirical Properties of Two Classes of Designs for Transferable Discharge Permit Markets," *Journal of Environmental Economics and Management* vol. 9, no. 2 (June 1982) p. 115; Scott E. Atkinson and T. H. Tietenberg, "Approaches for Reaching Ambient Standards in Non-Attainment Areas: Financial Burden and Efficiency Considerations," *Land Economics* vol. 60, no. 2 (May 1984) pp. 155 and 157, tables 1 and 2; Robert W. Hahn and Roger G. Noll, "Designing an Efficient Permits Market," in Glen R. Cass, et al., eds., *Implementing Tradeable Permits for Sulfur Oxide Emissions: A Case Study in the South Coast Air Basin, vol. II Main Text,* a report prepared for the California Air Resources Board by the Environmental Quality Laboratory of the California Institute of Technology (June 1982), pp. 106 and 110; Eugene P. Seskin, Robert J. Anderson, Jr., and Robert O. Reid, "An Empirical Analysis of Economic Strategies for Controlling Air Pollution," *Journal of Environmental Economics and Management* vol. 10, no. 2 (June 1983) p. 120; Alan J. Krupnick, "Costs of Alternative Policies for the Control of NO_2 in the Baltimore Air Quality Control Region," (unpublished Resources for the Future working paper, 1983) table 4, p. 22; Albert Mark McGartland, "Marketable Permit Systems for Air Pollution Control: An Empirical Study," (Ph.D. dissertation, University of Maryland, 1984) table 5.1, p. 74a; Walter O. Spofford, Jr.,

Table 12. (*Continued*)

Study	Air or water quality level[a]	Ratio of permit expenditures to market abatement cost		Ratio of total financial burden to CAC abatement cost	
		Ambient permit	Emission permit	Ambient permit	Emission permit
Krupnick (1983)					
Nitrogen dioxide	1	n.a.	2.00	n.a.	4.36
McGartland (1984)[c]					
Total suspended					
particulates	1	n.a.	1.14	n.a.	n.a.
	2	n.a.	1.33	n.a.	n.a.
	3	n.a.	1.51	0.66	n.a.
	4	n.a.	1.86	n.a.	n.a.
Spofford et al. (1984)[c]					
Sulfur dioxide	1[d]	n.a.	1.00	n.a.	2.39
Particulates	1[e]	n.a.	2.86	n.a.	0.35
Uniformly mixed assimilative pollutants:					
Harrison (1983)					
Noise	1	N.A.	2.84[f]	N.A.	n.a.
Uniformly mixed accumulative pollutants:					
Palmer et al. (1980)					
Chlorofluorocarbons	1	N.A.	15.40	N.A.	8.4

Clifford S. Russell, and Charles M. Paulsen, *Economic Properties of Alternative Source Control Policies: An Application to the Lower Delaware Valley* (Washington, D.C., Resources for the Future, 1984 manuscript in preparation) tables 4-20 and 5-20, pp. 4-102 and 5-100; David Harrison, Jr., "Case Study 1: The Regulation of Aircraft Noise," in Thomas C. Schelling, ed., *Incentives for Environmental Protection* (Cambridge, Mass., MIT Press, 1983) p. 131; Adele R. Palmer, William E. Mooz, Timothy H. Quinn, and Kathleen A. Wolf, *Economic Implications of Regulating Chlorofluorocarbon Emissions from Nonaerosol Applications,* Report #R-2524-EPA prepared for the U.S. Environmental Protection Agency by the Rand Corporation (June 1980) p. 131.

[a] A "1" designation implies the highest quality level considered by that study while a "2" signifies the second highest quality and so on.

[b] These numbers derived from Atkinson and Tietenberg (1984). They correspond to the cost of reaching the secondary standard.

[c] These estimates are based on a uniform emission tax rather than emission permits. Mathematically the two can be used interchangeably.

[d] Data for sulfur dioxide.

[e] Data for particulates.

[f] Ratio of payments for noise permits to other landing fees at Logan Airport in Boston, Mass.

Although this is an intriguing result, its policy significance is diminished by the complexities associated with an auction in ambient permits.[10]

In contrast to the results for ambient permits, financial burdens for emission permit auctions are frequently higher than the command-and-control financial burden. Five out of the eight air pollution studies represented show financial burdens to be higher for emissions permits. In part this is due to the higher control costs associated with the inherent overcontrol involved when using this approach to reach the ambient standards.

The tendency for emission permit auctions to lead to higher financial burdens than the command-and-control allocation is even more common in water pollution control than it is in air pollution control, despite the fact that several of these observations involve situations where the emission permit auction is cost effective. Every observation listed in table 13 found the emission permit auctions to be more expensive. High permit expenditures are the sole possible reason for the higher financial burden in these studies.

This conclusion is reinforced by a close examination of the first two data columns of tables 12 and 13. A ratio of 1.0 (or greater) means that permit expenditures are equal to (greater than) the corresponding abatement costs for the auction market. Ideally, permit expenditures would represent a small proportion of control costs, but, as both tables reveal, that is rarely the case. Even in the case with the smallest ratio (water quality level 2 in the Upper Hudson study), permit expenditures are 20 percent of control cost. Of the 39 numerical entries in the first two data columns of tables 12 and 13, permit expenditures are at least as large as control costs in 22 cases. In the case of chlorofluorocarbon regulation, permit expenditures would be over fifteen times larger than control costs. Large permit expenditures are the normal, rather than the exceptional, outcome.

This conclusion implies that while horizontal equity may be served by an auction, vertical equity is not. Though auction markets treat existing and new sources symmetrically, eliminating the new source bias, they also raise the cost of compliance substantially. When these higher costs get passed on in higher prices, the regressivity of the policy is increased.

The magnitude of the financial burden estimated by these studies is large enough to undermine the political feasibility of adopting a revenue auction market, particularly in a decentralized political system where the choice is up to the states. Any state unilaterally choosing an auction market would raise the financial burden for industries within its jurisdiction above those of competitors in jurisdictions not relying on an auction market. Not only would its existing industries be more vulnerable to

10. The interdependencies among the markets make the bidding process unrealistically complicated. See the discussion in chapter 4.

Table 13. The Size of the Potential Regional Financial Burden: Water Pollution

Study	Air or water quality level[a]	Ratio of permit expenditures to market abatement cost		Ratio of total financial burden to CAC abatement cost	
		Ambient permit	Emission permit	Ambient permit	Emission permit
Lyon (1982)	1[b]	n.a.	1.52	n.a.	2.02
Phosphorus and BOD	1[c]	n.a.	0.27	n.a.	1.10
Eheart, Brill, and Lyon (1983) BOD	Williamette River				
	1	n.a.	0.34	n.a.	1.12
	2	n.a.	0.22	n.a.	1.09
	Delaware River				
	1	n.a.	0.85	n.a.	1.18
	2	n.a.	1.01	n.a.	1.15
	3	n.a.	4.24	n.a.	2.60
	Mohawk River				
	1	n.a.	0.39	n.a.	1.20
	2	n.a.	0.30	n.a.	1.15
	Upper Hudson River				
	1	n.a.	0.48	n.a.	1.21
	2	n.a.	0.20	n.a.	1.09

Note: n.a. = Not available.

Sources: Randolph M. Lyon, "Auctions and Alternative Procedures for Allocating Pollution Rights," *Land Economics* vol. 58, no. 1 (February 1982) p. 24; J. Wayland Eheart, E. Downey Brill, Jr., and Randolph M. Lyon, "Transferable Discharge Permits for Control of BOD: An Overview," in Erhard F. Joeres and Martin H. David, eds., *Buying a Better Environment: Cost-Effective Regulation through Permit Trading* (Madison, Wis., University of Wisconsin Press, 1983) table 1, p. 177.

[a] A "1" designation implies the highest quality level considered by that study while a "2" signifies the second highest quality and so on.
[b] Data for phosphorus removal.
[c] Data for BOD removal.

closures, but new industries would find that state a relatively unattractive place to locate. Since no state wants to put itself in this position if it can be avoided, auction markets are not politically feasible.

The Distribution of Source Financial Burden

The preceding examination of regional financial burden sheds a considerable amount of light on whether the *average* source is better off under the auction system or command-and-control approach as well as on the *average* size of the permit expenditures relative to the control costs.

However, it also masks a considerable amount of variability among sources. Some sources will bear a lower financial burden than the average and others will bear more. Among other reasons, these differences are important for industrial polluters because unequal cost increases may cause shifts in market shares.

How serious is this unequal distribution of the burden? In his study of the control of aircraft noise, Harrison (1983, p. 114) concludes that the extra payments to the government involved in an auction market would create greater cost disparities among airlines than the traditional approach.[11] He also notes, however, that the disparities would not be particularly great among close competitors. Cargo airlines, which typically operate the noisiest aircraft at night, when the sensitivity to noise is the highest, would purchase the most permits. However, cargo airlines operating similar planes on similar rates would face similar cost increases; little competitive edge would be gained or lost by direct competitors as a result of these payments. He speculates, but does not perform the calculations to test his speculation, that the variability among passenger airlines might be a more serious problem.

Palmer et al. (1980) also examine the interindustry and interfirm effects of using an auction market to control chlorofluorocarbons (figure 5). Though this particular example is somewhat unusual because of the unusually large permit expenditures that would be involved, the variability of costs introduced by an auction market is striking.

One of the aspects of this figure that is particularly noteworthy is the "other" category. This category contains some product areas (rigid insulating foams, liquid fast freezing, and sterilants) where the authors found that no control methods would be introduced, even in an auction market. For these product areas, the *only* expense is incurred in purchasing permits and it represents a very large outlay. The manufacturers of these product lines could be expected to be unusually vociferous in their opposition to an auction approach and unusually warm in their support of a traditional regulatory approach which, in all likelihood, would place no control requirements on them at all.

The tendency for auction markets to increase the cost disparities is not perfectly general. Lyon (1982) shows that firms with low marginal control costs will prefer uniform command-and-control regulations, while firms with high marginal control costs may or may not prefer auction markets. If in an auction market the control cost savings to the high-cost source are large enough, its cost disadvantage to the low marginal cost source would be reduced relative to the command-and-control approach in any move from command-and-control to an auction approach.

11. Harrison bases this conclusion on a noise charge rather than an auction market. Given the mathematical equivalence of the two approaches, the conclusion holds for auction markets as well.

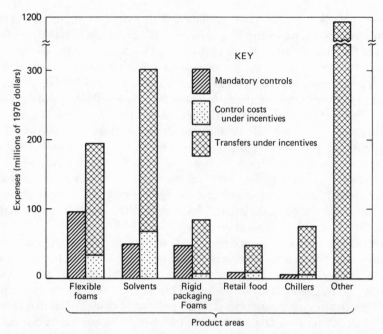

Figure 5. Cumulative industry expenses under mandatory controls and an auction market for permits

The discussion so far has focused on the distribution of financial burden among sources when the location of the sources is not a factor. The results of the Harrison study, for example, were in terms of a uniform marginal cost of control for each effective perceived noise decibel[12] whereas the Palmer et al. study is based on a uniform marginal cost of controlling chlorofluorocarbon use in the United States.

When location is considered, either by introducing an auction market for ambient permits or by establishing different emission permit auction markets in different local areas, another source of cost variability is introduced. Sources in geographic areas requiring higher marginal control costs would face higher financial burdens than direct competitors in areas with lower marginal control costs simply by virtue of their location.

Existing sources are particularly vulnerable because their location would have been determined before the system was implemented. Unlike new sources, they could not pick a location to minimize costs. Although this cost differential would have in all likelihood been unanticipated by the existing sources, it could have a major impact on their

12. Aircraft noise is universally measured in terms of effective perceived noise decibels. This scale modifies the basic decibel scale by accounting for how people judge the noisiness of aircraft takeoffs and landings. High-frequency noises, for example, receive more weight.

market shares. Short of moving, there is little existing sources can do to unburden themselves of this competitive disadvantage. This is the flip side of the location coin. While considering the location of the emissions provides the proper incentives for *new* sources to locate in low damage areas, it also imposes higher costs on those sources which have already located in what turns out after the fact to be a high marginal control cost area, diminishing their competitiveness in the process.

One way to alleviate these unfortunate side effects of auctions is to eliminate the transfer to the government either by grandfathering or by using a zero-revenue auction.[13] Since for a given distribution rule and set of credit prices these two approaches would yield the same distribution of financial burden among sources, we discuss only the grandfathering results. It should be understood, however, that these results apply with equal force to the zero-revenue rule.

Grandfathering the ultimate financial burden among sources is contingent upon the distribution rule used to define the baseline control responsibility. One of the earliest studies of the distribution of financial burden in the grandfathering context (David et al., 1980, table 1) examined the control of phosphorus effluent in Lake Michigan. Simulating an initial distribution of control responsibility to 53 plants based on their average daily flow, the data in this study support two interesting conclusions: (1) three sources received revenue from the sale of emission reduction credits that was large enough to more than cover their control costs, leaving them with a profit, and (2) these same three sources plus a fourth sold 95 percent of all traded credits.

Both of these results were caused by economies of scale. All four of the major credit sellers had the largest average waste flow. An initial distribution rule that allocates baseline control responsibility in proportion to the flow of waste fails to account for the fact that, due to economies of scale, large sources frequently can control a larger percentage of waste at a lower marginal cost than can smaller sources. As a result, these sources frequently can sell a relatively large number of credits. Distribution rules based on some proportion of waste flow provide relatively more compensation to large sources.

This underallocation of control responsibility sets the stage for a second problem. Because of their large excess supply of credits, a few large sources could dominate the market, using their influence to command higher than normal profits. Since this topic is discussed in some detail in

13. Another is to use an incentive-compatible auction rather than the single-price auction we have considered. Designed primarily as a means of preventing price manipulation, this auction involves lower permit prices. Since Lyon (1982, pp. 23–25) has shown that in two different cases the reductions in permit expenditures are not dramatic, we reserve our discussion of this approach for chapter 6.

the next chapter, for this chapter we merely point out that initial distribution rules based on some proportion of waste flow raise the specter of creating an imperfect market.

Few studies have looked at the distributional costs associated with grandfathering a concentration credit system. One that has (Spofford et al., 1984, table 4-21, pp. 4-104 and 4-105) examines the effects of a proportional distribution rule on the compliance costs of several large categories of sources in the Lower Delaware Valley as they reduce sulfur dioxide to meet the ambient standard. Spofford's results reinforce the conclusion reached when location was ignored—very large sources sell permits. In fact, in this study electrical generating plants and other large air pollution sources made enough from the sale of the permits to completely finance their pollution control program. In contrast, petroleum refineries were estimated to have their compliance costs substantially increased by the need to purchase additional permits.

Although it is difficult to defend any particular distribution rule as best, political feasibility suggests using the command-and-control allocation to define the baseline control responsibility. As long as the command-and-control allocation is compatible with reaching attainment, this allocation would allow a smooth transition into the emissions trading program.

Not only would this particular distribution rule minimize any transitional friction by explicitly linking it to the previous policy, but every existing source would be at least as well off as it would have been in the command-and-control approach. Many would be better off. Both individually and collectively, existing sources would prefer a permit approach using this distribution rule to the command-and-control approach because no source could be worse off than it was under command-and-control. By refusing to trade, the source could always retain its command-and-control responsibility when that was preferred. To the extent that existing sources are able to mount effective political opposition, as long as the command-and-control allocation already exists, this distribution rule would produce the least resistance to change.

The property that no source is harmed by the introduction of the emissions trading system does not necessarily hold in areas where the command-and-control allocations are not sufficient to meet the ambient standards. Whenever the emission trading system is introduced as a means of improving air quality, additional control would be necessary, at least at some locations. Since additional control would be needed, there can be no guarantee that at least one source would not receive a larger financial burden as the air quality is improved.

One study (Atkinson and Tietenberg, 1984) has examined the possibilities in St. Louis of using emissions trading to reach attainment while being sensitive to the resulting distribution of financial burden. The first finding was that the initial command-and-control allocation (which resulted in air quality violating the ambient standards) involved such excessive control costs that it was possible to lower regional control costs while improving the air quality sufficiently to meet the primary ambient standard for particulates. By taking source location into account, it was possible to construct a distribution rule that produced better air quality at the binding receptors, but required less emission reduction than the command-and-control allocation. According to this distribution rule, those sources having the largest impact on the receptors of interest increased their control, while those having the smallest impact decreased their control by an even larger amount.

While this approach clearly holds down regional control cost, it concentrates the additional control responsibility on a few sources. This could pose quite a dilemma for any control authority, particularly in a command-and-control approach without emissions trading. Since concentrating the financial burden on a few sources could well appear discriminatory, provoking political and legal opposition, the control authority would probably be forced to allocate the responsibility among all sources to create the appearance, if not the reality, of evenhanded treatment. Cost effectiveness would be sacrificed to preserve an appearance.

Due to the potential for cost sharing, the issue of which sources bear the financial burden can be separated from the issue of which sources undertake the control in emissions trading programs. Both the appearance of propriety and cost effectiveness can be preserved. The initial responsibility for additional control could be allocated across all sources, giving them all a share of the financial burden. Those sources that could control most cheaply would sell concentration reduction credits and those facing very high marginal control costs would purchase permits.[14] Although all sources would participate in the financing of the additional control, the degree of control and the particular sources adopting the more stringent control technologies would be cost effective. The combination of cost effectiveness and cost sharing available with emissions trading allows the ambient standards to be achieved in nonattainment areas with few if any sources facing a financial burden greater than that faced under a command-and-control strategy which failed to achieve the standards. Cost sharing is one of the powerful properties of emissions trading that is forfeited by the command-and-control approach.

14. This approach presumes the use of a pollution offset trading rule. The proportional initial allocation of control responsibility would require much larger than necessary emission reductions. These would significantly constrain trading opportunities if either the nondegradation or modified pollution offset rules were used.

The potential for cost sharing is apparently very large. One study in the Southwest found that current regulations control emissions from power plants at a marginal cost of 18 to 25 cents per pound of SO_2 while copper smelters in the same area could reduce their emissions by 60 percent at a marginal cost of 0.6 cents per pound.[15] Because copper smelting is a financially precarious industry, it has escaped controls in the past, leaving this low-cost source of emission reduction untapped. Emissions trading makes it possible for power plants to finance reductions at the smelters which would be cheaper than reducing their own emissions to a higher degree.

THE EPA EMISSIONS TRADING PROGRAM

Although we have identified several possible approaches for distributing the financial burden among sources, it is probably not surprising that the EPA emissions trading program is a version of grandfathering. The command-and-control responsibility established by state implementation plans has been used to define the control responsibility baseline for all sources to use in determining their interest in the acquisition or sale of emission reduction credits.

Though the emission reduction credit program is similar to grandfathering based on an initial distribution rule that is determined by command-and-control, there are important differences. One of the most important of these involves restrictions on the certification and use of emission reduction credits. Not all credits may be certified, nor all certified credits transferred to all interested parties. These restrictions serve to lessen the cost effectiveness of the policy as well as to impede progress toward attainment.

The Baseline Issue

One condition that must be satisfied for any emission reduction to be certified as a credit is for the reduction to be surplus.[16] To be considered surplus, the reduction must be over and above some baseline reduction required by the control authorities in the state implementation plans. Though the application of this requirement would appear to be rather straightforward, in practice that has not been the case.

For the purposes of examining the baseline issue, it is important to distinguish between those air quality control regions that have approved

15. See Kneese and Brown (1981, pp. 30–31).
16. Emissions Trading Policy Statement, 47 *Federal Register* 15076 at 15086 (April 7, 1982).

plans demonstrating current or eventual attainment and those with no such demonstrations. In those areas with approved plans, the problem is that there are two competing baseline definitions—allowable emissions and actual emissions. In areas without approved plans, the problem is the lack of any acceptable definition.

To understand how two baselines could be possible in areas demonstrating attainment, it is necessary to delve a bit into the mechanics of how the command-and-control emission standards which form the basis for the baseline were established. Emission standards are typically developed for categories of sources, not tailored to individual sources. In order to develop standards applicable to all facilities within the category, the control authority must establish emission limits at levels which reflect performance under the worst, reasonably-to-be-expected situation. To do otherwise would establish standards that were not consistently achievable by all sources.

In practice, the actual operating emission rates from many of these facilities fall well below the allowable limits. For example, the new source performance standard for Portland cement plant kilns was set at 0.15 kg/mg of dry feed. Initial compliance test results for 29 kilns subject to the standard averaged 0.073 kg/mg (less than half the allowable emissions). For the average plant, the difference between allowable and actual emissions is about 50 tons of particulate emissions per year.[17]

As environmentalists have pointed out, using an allowable emissions baseline in this circumstance can result in trades which increase emissions. Any source purchasing an emission reduction credit based on a reduction in *allowable* emissions (but no reduction in actual emissions) could use that credit to increase *actual* emissions. As far as the SIP is concerned, however, there would be no compensating actual decrease because the difference between allowable and actual emissions was already accounted for in developing the compliance schedule. Though on paper this trade would hold allowable emissions constant, in fact actual emissions would increase.

To prevent this double counting of emission reductions, the EPA policy guidance requires the basis for defining the baseline for credits in nonattainment areas to be consistent with the basis used for demonstrating attainment in the SIP. If the state used actual emissions in constructing the SIP, then actual emissions should be used as the basis for determining whether any given reduction is surplus. Otherwise, the baseline could be defined in terms of allowable emissions.

The guidance is successful in resolving a potential inconsistency between the SIP and the baseline for the emission reduction credit because it ensures that trades are consistent with the attainment demonstration.

17. This information was taken from an internal EPA memorandum proposing an amendment to the NSPS standards for the Central Illinois Public Service Company.

As long as allowable emissions can be shown to be compatible with attainment of the ambient standards, there is no double counting of these reductions.

In achieving this consistency between each state's plan and its baseline, however, this resolution has introduced an inconsistency in how sources are treated among the various states. Two otherwise identical sources in different nonattainment areas making identical control technology choices would not necessarily be treated the same. Sources in nonattainment regions where the SIP was based upon allowable emissions could claim a certified emission reduction credit, while sources in nonattainment regions where the SIP is based on actual emissions could claim no such credit. This difference arises not out of any grand design, but out of the need to respond to a historical accident.

Fortunately this is a transitory phenomenon arising out of a need to shoehorn the emissions trading program into a policy shaped by past control decisions. Future decisions made by sources need not trigger similar concerns. Agreements with the control authorities in the future can automatically stipulate either allowable emissions or actual emissions as the baseline for both the SIP and the definition of the emission reduction credit. As long as they are consistently defined, a workable program can be maintained.

In areas without approved plans demonstrating attainment, the issue is less tractable. For these areas, in the absence of a demonstration to the contrary, it cannot be assumed that either an allowable or actual emission baseline would be consistent with attainment. It would be dangerous to allow trading in those areas prior to the establishment of an approved baseline. Because a tightening of the existing baseline would be necessary, sources have an incentive to anticipate this tightening by overcontrolling with respect to the old baseline and selling the excess reduction to another source. Instead of waiting to be required to control more when the new baseline is imposed, the source gains by jumping the gun. The extra control which would have ultimately been required anyway is sold for a profit. The end result is that trades consummated before the revised baseline is imposed (and to some extent in anticipation of it) make it more difficult to revise the baseline upward as needed. Consummated trades create a legal foundation for resisting subsequent more stringent requirements.

The emissions trading policy here is less satisfactory. In some cases an actual emissions baseline is acceptable whether or not that baseline is consistent with attainment.[18] While this approach facilitates trading, it does so by possibly delaying eventual attainment.

18. 47 FR 15079 at 15080 (April 7, 1972). EPA is currently reexamining this policy. See 48 FR 39580 (August 31, 1983).

Shutdown Credits

The treatment of emission reductions gained from plant closures is another issue which has been embroiled in a considerable amount of political turmoil. It is a significant issue because shutdowns are frequently the most important source of emission reductions in nonattainment areas. The dictates of cost effectiveness are quite clear on how shutdown credits should be treated. Shutdowns are a legitimate and significant source of emission reductions which should become integrated into the quest for better air quality at lower cost. The failure to provide property rights for shutdowns provides incentives to retain uneconomic capital equipment longer, delaying the receipt of the resulting emission reduction.

Current policy is not completely in accord with these dictates. For the offset policy, the extent to which shutdown credits can be used as offsets is limited. Oregon, for example, has adopted a "use it or lose it" policy toward shutdowns.[19] Permanent source shutdowns cannot be banked under Oregon rules. The credits can either be traded within one year after the permanent shutdown or can be incorporated into a plan for the future growth of the plant. Growth plans must be approved within one year of the shutdown. If the credits are neither traded nor reserved in an approved growth plan within the one-year period, their ownership reverts to the state to be used either to improve air quality or to be allocated to new sources.

Generally in the bubble policy, EPA allows the source to claim a shutdown credit in nonattainment regions as long as the reductions were not previously counted in the states' plans for meeting the reasonable further progress requirement. As their intended means of fulfilling the reasonable further progress requirement, many states had estimated that a certain number of shutdowns would occur, using the resulting emission reductions as their means of improving air quality. To allow these credits to be used in bubble trades would prevent their use in satisfying the reasonable further progress requirement and jeopardize the compliance schedule.

One interesting characteristic of this approach is that EPA applies it only in nonattainment areas; shutdown credits can generally be used in bubble trades once attainment has been reached. This creates a perverse incentive for marginal plants. To maximize their own wealth, they would delay closure until after attainment has been achieved (to the extent that is possible) because only sources which shut down after attainment is reached can qualify for the credit. Thus a rule which is designed to hasten compliance probably ends up actually delaying it by deferring the impact of a potentially large source of emission reductions.

19. See Kostow and Kowalcyzk (1983, p. 984) for a brief description of the Oregon approach.

The netting portion of the emissions trading program was introduced by EPA in part to provide a means by which major sources undergoing modification or expansion could escape this bias. When the emissions from one of these sources could be netted out by an internal offset (frequently a curtailment or shutdown credit), the new source requirements could be avoided. In essence this created a limited opportunity for modifying plants to escape the LAER requirement.

In practice, netting has had a limited impact. Even when unencumbered by judicial restraints, the program benefits only those modified sources having internal offsets available to them. New sources locating in an area for the first time would be granted no relief. And though ultimately supporting the program, the court initially took a dim view of using the netting program in nonattainment areas, seeing it as a threat to improving air quality.[20]

Minimum Control Thresholds

One source of cost ineffectiveness embedded in the early versions of the program which has persisted even in more recent versions is the minimum control thresholds imposed on new or major modified sources. A summary of these requirements is given in figure 6. The last column was added by the netting portion of the program.

	Existing sources	Major new or modified sources	Major modifications without significant emission increases
Attainment areas	None	Best available control technology	New source performance standards
Nonattainment areas	Reasonably available control technologies	Lowest achievable emission rate	New source performance standards

Figure 6. Stationary source statutory performance standards

20. See *Natural Resources Defense Council, Inc.* v. *Gorsuch* (August 17, 1982). This decision was overruled by the U.S. Supreme Court in *Chevron U.S.A.* v. *Natural Resources Defense Council Inc.* 52 LW 4845 (June 25, 1984).

In a cost-effective allocation, these emission standards would establish a baseline control responsibility, with sources encouraged to comply with that target in the cheapest way possible, *including the acquisition of either internal or external emission reduction credits.* The current emissions trading program forbids the use of acquired emission reduction credits as a means of meeting the LAER, BACT, or NSPS standards. Only the RACT standards can be satisfied with emission reduction credits. Compliance with the other standards can only be demonstrated by achieving the emission reductions designated for each emission point.

The effect of a minimum treatment standard such as LAER is depicted in figure 7. In that figure, the situation for two possible trading sources is given. Initially, by assumption, the new source is not operating so its allowable emissions are zero. Any subsequent emission levels from the new source greater than E_1^2 are disallowed by the LAER standard. The allowable emissions for the existing source are assumed to be E_1^0.

The LAER standard reduces the trading possibilities and increases the cost of compliance for new sources. If the LAER standard were not in effect, the trading possibilities would be described by areas $A + B + C$. Any emission-rate pair within those areas would satisfy the air quality standards. With the LAER standard, the trading possibilities are reduced to area A. (Area B is still feasible, but because the new source gains nothing from any further control by the existing source beyond E_1^1, it would not purchase those further reductions.)

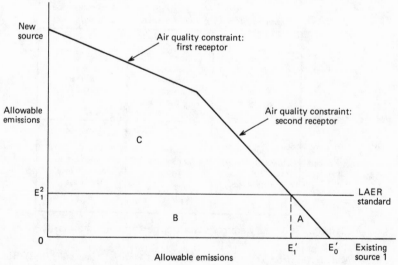

Figure 7. The effect of a minimum treatment standard

Coupled with the restrictions on the use of shutdown credits as offsets and the necessity for acquiring offsets for any remaining emissions, this minimum control requirement makes the new source bias substantially larger than it need be. New sources face higher than necessary costs not only because they cannot use emission reduction credits acquired from other sources to meet the LAER, BACT, and NSPS standards, but also because the restrictions on the use of shutdown credits as offsets reduce the supply of available offsets and increase the price of those that remain. In addition, only new sources have to acquire emission reduction credits to cover all uncontrolled emissions.

This new source bias causes the turnover of capital equipment to occur more slowly than would otherwise be the case. Older, more heavily polluting processes are retained longer while newer, less heavily polluting processes wait on the sidelines. By extending the useful economic life of the most heavily polluting plants, the current emissions trading program is making the attainment of the ambient standards more costly and less rapid.

The Roots of the Problem

Though these problems have been dealt with by the EPA on an *ad hoc* basis, it is important to realize that they stem from a common cause—the failure to develop the air quality accounting system necessary to make the system work effectively. This accounting system is essential if the emissions allowed in operating permits are to be completely consistent with progress toward attainment in nonattainment areas and with the orderly consumption of the allowable increment in PSD regions. Only by accounting for all emissions in an accurate emissions inventory and relating this inventory to the attainment of the standards can emission trades be confidently seen as one avenue to attainment.

Contrast this with the current approach. The ambiguity and uncertainty over the determination of whether emission reduction credits are surplus or not results from the program's historical focus on emission reductions from a predefined baseline rather than on a baseline defined in terms of the ambient standards. Because the emissions trading baseline was determined prior to the inception of the emissions trading program, the EPA was forced to allow different approaches in different states. The potential inconsistency between the introduction of the emissions trading program and the achievement of the ambient standards was solved on an *ad hoc* basis.

Because the requirement that nonattainment areas demonstrate reasonable further progress has historically been defined in terms of emissions rather than air quality, the connection between the emissions trading program and achieving the ambient standards has been necessarily loose. Annual reductions in emissions are neither necessary nor sufficient for improvements in air quality at those monitors recording pollution levels exceeding the standards.

With the achievement of the ambient standards in doubt due to the absence of an adequate air quality accounting system, the states have followed an excessively risk-averse strategy. Immediate extra reductions were sought wherever they could be found in the hope they would be sufficient. As part of this frenzied search for reductions, shutdown credits have been confiscated, preventing their use for any other purpose and minimum control thresholds have been imposed to prevent emission reduction credits being used to satisfy BACT, LAER, and NSPS standards. Allowable emissions trades that permit actual emission increases are also an outgrowth of an inadequate air quality accounting system.

The ability to develop an accurate and effective air quality accounting system hinges on the ability of state control authorities to define a unique allowable emissions level for each major source, to issue permits consistent with that definition, and to monitor for compliance. These do not seem insurmountable problems, particularly since Oregon already has the basic ingredients of this approach in place.[21] At the core of the Oregon rules lies the plant site emission limit, an allowable emissions standard placed on all new and existing sources larger than some minimum emissions threshold. These emission limits are incorporated into the operational permits received by the sources and are based on actual emissions in 1977 or 1978 minus any required controls since that time. The total amount of pollution allowed by these limits is consistent (including some anticipated reversion of shutdown credits to the state) with the baseline in attainment areas. Any emission trades result in adjustments in the operating permits of both the acquiring and relinquishing sources. With this type of air quality accounting system, emission trades can be conducted without jeopardizing air quality. Without an adequate air quality accounting system, states will either continue to engage in cost-ineffective practices to protect air quality or air quality can deteriorate. Neither alternative is attractive.

SUMMARY

• By the manner in which it assigned the control responsibility, the command-and-control approach imposed a very large and unequally distributed financial burden on complying sources. Because they bore a

21. Kostow and Kowalcyzk (1983, p. 982).

disproportionate share of the control responsibility, new sources and expanded or modified existing sources faced particularly high costs of compliance. In part due to this new source bias, the job losses from closures of existing plants have been very small.

• The ultimate incidence of the control costs associated with the command-and-control approach has particularly penalized the poor. The control costs associated with the traditional approach were substantially higher than needed, causing unnecessarily large increases in industrial prices to cover these costs. Because the poor use a higher proportion of their income to purchase these commodities, these price increases consume a larger proportion of the budgets of poor households than wealthier households.

• Because under certain conditions the allocation of the baseline control responsibility can be used to affect the distribution of costs without increasing the amount spent on control technologies, emissions trading offers the opportunity to achieve both cost effectiveness and fairness. The absence of a tradeoff between these objectives is quite unusual; this is a unique opportunity.

• The control authority has many options in how it distributes the financial burden. It can auction off allowable emission or concentration permits, it can purchase allowable emission or concentration permits from the sources, or it can allocate a baseline control responsibility on the basis of some initial distribution rule, letting the ultimate allocation of emission or concentration reduction credits be determined by the market.

• From the point of view of the source required to control its emissions, there are two components of financial burden: (1) control costs and (2) credit or permit expenditures. The empirical evidence suggests that the latter frequently would be larger in magnitude than the former if an auction system were to be used. Permit expenditures are sufficiently large that sources would typically have lower financial burdens under the traditional approach than an auction permit approach. This significant source of political opposition to auction markets undermines its acceptance.

• Because it would inevitably require large increases in the public budget and because it creates perverse location incentives, subsidization is not a politically popular option either.

• Grandfathering and zero-revenue auctions provide two ways to allocate the control responsibility without financial transfers to or from the government. Both require the control authority to specify some rule for distributing the baseline control responsibility.

• Because they ignore economies of scale, rules proportionately allocating control responsibility among sources on the basis of some proportion of waste flow usually create an after-market where a few large sources sell credits.

• A distribution rule using the administratively determined command-and-control allocation of control responsibility to provide the starting point for emission trades has the property that no existing source is worse off with emissions trading than it was under the command-and-control allocation. Sources consummating trades are better off, while nontrading sources are no worse off than they would have been had the trading system not been established.

• One of the more significant differences between a command-and-control approach and transferable permit approach is the degree to which they allow cost sharing. Cost sharing is possible with transferable permits, but it is not possible with command-and-control regulation.

• Cost sharing is valuable as a way of separating the decision of what emission points are to be controlled from the decision as to how reductions will be paid for. By using the cost-sharing possibilities in emission trading, it is feasible to secure low-cost emission reductions from financially precarious firms by allowing them to sell emission reduction credits. Control authorities using command-and-control regulations would normally have to bypass these sources, leaving more control on other sources to make up the difference.

• Restrictions on the use of shutdown credits and minimum control thresholds are two ways in which the EPA emissions trading program deviates from cost effectiveness. Both of these characteristics tend to force larger-than-necessary costs on new sources. By extending the economic life of existing plants (which tend to be more heavily polluting than new plants), these restrictions are making the attainment of the ambient standards more costly and less rapid than necessary.

• These restrictions as well as the possibilities for trades which increase emissions result from the failure to develop and to implement a reliable air quality accounting system which ties operating source permits to reasonable further progress in nonattainment areas and to the consumption of the increment in PSD regions. In the absence of this system, statutory requirements for states to demonstrate attainment have led to myopic strategies which sacrifice incentives for future emission reduction in order to gain as much short-term reduction as possible. Some states have begun to implement the requisite kind of accounting system, so it is clearly possible.

REFERENCES

Atkinson, Scott E., and T. H. Tietenberg. 1982. "The Empirical Properties of Two Classes of Designs for Transferable Discharge Permit Markets," *Journal of Environmental Economics and Management* vol. 9, no. 2 (June) pp. 101–121.

———, and ———. 1984. "Approaches for Reaching Ambient Standards in Non-Attainment Areas: Financial Burden and Efficiency Considerations," *Land Economics* vol. 60, no. 2 (May) pp. 148–159.

David, M., W. Eheart, E. Joeres, and E. David. 1980. "Marketable Permits for the Control of Phosphorus Effluent into Lake Michigan," *Water Resources Research* vol. 16, no. 2 (April) pp. 263–270.

Dorfman, Nancy S., and Arthur Snow. 1975. "Who Will Pay for Pollution Control? The Distribution by Income of the Burden of the National Environmental Protection Program, 1972–80," *National Tax Journal* vol. 28, no. 1 (March) pp. 101–115.

Eheart, J. Wayland, E. Downey Brill, Jr., and Randolph M. Lyon. 1983. "Transferable Discharge Permits for Control of BOD: An Overview," in Erhard F. Joeres and Martin H. David, eds., *Buying a Better Environment: Cost-Effective Regulation Through Permit Trading* (Madison, Wis., University of Wisconsin Press) pp. 163–195.

Gianessi, Leonard P., Henry M. Peskin, and Edward Wolff. 1979. "The Distributional Effects of Uniform Air Pollution Policy in the United States," *Quarterly Journal of Economics* vol. 93, no. 2 (May) pp. 281–301.

Gianessi, Leonard P., and Henry M. Peskin. 1980. "The Distribution of the Costs of Federal Water Pollution Control Policy," *Land Economics* vol. 56, no. 1 (February) pp. 85–102.

Hahn, Robert W., and Roger G. Noll. 1982. "Designing an Efficient Permits Market," in Glen R. Cass, et al., eds., *Implementing Tradeable Permits for Sulfur Oxide Emissions: A Case Study in the South Coast Air Basin,* vol. II Main Text, a report prepared for the California Air Resources Board by the Environmental Quality Laboratory of the California Institute of Technology (June) pp. 102–134.

Harrison, David, Jr. 1983. "Case Study 1: The Regulation of Aircraft Noise," in Thomas C. Schelling, ed., *Incentives for Environmental Protection* (Cambridge, Mass., MIT Press) pp. 41–143.

———, and Paul R. Portney. 1982. "Who Loses from Reform of Environmental Regulation," in Wesley A. Magat, ed., *Reform of Environmental Regulation* (Cambridge, Mass., Ballinger) pp. 147–179.

Kneese, Allen V., and F. Lee Brown. 1981. *The Southwest Under Stress: National Resource Development Issues in a Regional Setting* (Baltimore, Md., Johns Hopkins University Press for Resources for the Future).

Koch, James C., and Robert E. Leone. 1979. "The Clean Water Act: Unexpected Impacts on Industry," *Harvard Environmental Law Review* vol. 3 (May) pp. 84–111.

Kostow, Lloyd P., and John F. Kowalcyzk. 1983. "A Practical Emission Trading Program," *Journal of the Air Pollution Control Association* vol. 33, no. 10 (October) pp. 982–984.

Lyon, Randolph M. 1982. "Auctions and Alternative Procedures for Allocating Pollution Rights," *Land Economics* vol. 58, no. 1 (February) pp. 16–32.

Maloney, Michael T., and Robert E. McCormick. 1982. "A Positive Theory of Environmental Quality Regulations," *Journal of Law and Economics* vol. 24, no. 1 (April) pp. 99–123.

McGartland, Albert Mark. 1984. "Marketable Permit Systems for Air Pollution Control: An Empirical Study" (unpublished Ph.D. dissertation, University of Maryland).

Montgomery, W. David. 1972. "Markets in Licenses and Efficient Pollution Control Programs," *Journal of Economic Theory* vol. 5, no. 3 (1972) pp. 395–418.

Palmer, Adele R., William E. Mooz, Timothy H. Quinn, and Kathleen A. Wolf. 1980. *Economic Implications of Regulating Chlorofluorocarbon Emissions from Nonaerosol Applications*, Report #R-2524-EPA prepared for the U.S. Environmental Protection Agency by the Rand Corporation (June).

Portney, Paul R. 1981. "The Macroeconomic Impact of Federal Environmental Regulations," in Henry M. Peskin, Paul R. Portney, and Allen V. Kneese, eds., *Environmental Regulation and the U.S. Economy* (Baltimore, Md., Johns Hopkins University Press for Resources for the Future).

Seskin, Eugene P., Robert J. Anderson, Jr., and Robert O. Reid. 1983. "An Empirical Analysis of Economic Strategies for Controlling Air Pollution," *Journal of Environmental Economics and Management* vol. 10, no. 2 (June) pp. 112–124.

Spofford, Walter O., Jr., Clifford S. Russell, and Charles M. Paulsen. 1984. *Economic Properties of Alternative Source Control Policies: An Application to the Lower Delaware Valley* (Washington, D.C., Resources for the Future, manuscript in preparation).

U.S. Environmental Protection Agency. 1982. *Third Quarter Report of the Economic Dislocation Early Warning System* (Washington, D.C., Environmental Protection Agency).

ADDITIONAL READINGS

Brady, Gordon L., and Richard E. Morrison. *Emission Trading: An Overview of the EPA Policy Statement*. Policy and Research Analysis Report 82-2 (Washington, D.C., National Science Foundation, 1982).

Crandall, Robert W. *Controlling Industrial Pollution: The Economics and Politics of Clean Air* (Washington, D.C., Brookings Institution, 1983).

Quinn, Timothy H. "Distributive Consequences and Political Concerns: On the Design of Feasible Market Mechanisms for Environmental Control," in Erhard F. Joeres and Martin H. David, eds., *Buying a Better Environment: Cost-Effective Regulation Through Permit Trading* (Madison, Wis., University of Wisconsin Press, 1982) pp. 39–54.

Rose, Marshall. "Market Problems in the Distribution of Emission Rights," *Water Resources Research* vol. 9, no. 5 (October 1973) pp. 1132–1144.

6 / Market Power

The two previous chapters considered issues that not only have been recognized by the control authorities, but have had specific procedures developed to deal with them. Though these procedures have, on occasion, been somewhat crude, the fact remains that the current reform package has begun the process of coping with these particular challenges to the design of a smoothly running emissions trading program.

In contrast, neither EPA nor the states seem to have given much thought to the potential for market power to disrupt emissions trading. The rules governing the program do not reflect any particular concern with market power. If anything, the current rules increase, rather than reduce, the potential for market power.

Is this lack of concern justified? Under what circumstances can market power arise? What are the consequences of market power? Is it a serious problem? Does it threaten the ability of the program to meet its objectives? Can steps be taken to mitigate adverse consequences when necessary?

In the search for answers to these questions, we shall consider two rather different types of market power. The first type occurs when an aggressive source seeks to reduce its expenditure on emission credits or permits by manipulating the price. What happens to the other sources in this type of market power is incidental, not central; power over the permit market is an end in itself.

In the second type, a source or coalition of sources would attempt to use power in the permit or credit market as leverage in gaining power in product markets. By reducing the influence of its competitors, a predatory source would seek to increase its market share. Permit market power would be sought, not as an end in itself, but as a means of exerting influence over other markets.

It is necessary to differentiate these types of objectives for market power because each arises from a unique set of circumstances, has somewhat different consequences, and offers a somewhat different menu of mitigation possibilities. Therefore this chapter examines the evidence for each type, beginning with price manipulation.

PERMIT PRICE MANIPULATION

Suppose that one or more firms (which we shall call *price-setting* firms) seek to exercise control over credit or permit prices to reduce their financial burden. The extent to which they can accomplish that objective depends on a number of factors, one of which is the method chosen to govern the initial distribution of control responsibility.

Auctions and Subsidies

If auction markets were chosen to distribute permits, all sources would be seeking permits and the control authority would be the only seller. The ability of any source, or coalition of sources, to affect the prices paid for those permits would depend on the magnitude of their demand compared with the demands of the other sources. In the limit, a market with only one source, the source could acquire the permits at no cost. Since a bid of even zero dollars would be decisive, permit expenditures would be zero for this source, even in an auction.

When there are other sources, the situation is somewhat more complex. Suppose that one particular firm wished to exercise control over price while all others were content to act as price takers. The price-setting firm, by articulating an artificially low demand for permits, could lower the price.

The effects of this behavior on permit markets can be illustrated through the use of a simple numerical example. Not only does this example lend a certain concreteness to the argument, it turns out that many of the conclusions derived from it persist in more realistic situations.

Suppose a particular auction market was established by the government to auction off permits. The revenue from the sale of these permits is presumed to be retained by the government. For simplicity assume that only two sources are bidding for these permits. The first source is presumed to use its purchasing behavior to control the price while the second source is presumed to be a price taker. The first source adjusts its permit demands so as to minimize its financial burden, taking into account the reactions of the second source. In particular, it knows that the second source, being a cost minimizer, will choose that number of permits equating its marginal control cost to the price. The lower the price, the more permits it would acquire.

Let the marginal costs of control for each of the two sources be $MC_1 = Q_1$ and $MC_2 = 2Q_2$ where Q_1 and Q_2 are, respectively, the emissions reduced by the first and second source. In the absence of any control, these two sources are assumed to emit five units of emissions each (for a total of ten). Since the control authority is presumed to allow only four units, six units must be reduced. By auctioning off four permits, each worth one unit of emissions, the control authority could fulfill this objective.

By definition, the financial burden of the first (price-setting) source is the sum of its permit expenditures and control costs. Symbolically this source's decision to minimize its financial burden can be characterized as

$$\min_{Q_1} P(5 - Q_1) + Q_1^2/2 \qquad (1)$$

where the first term is permit expenditures (permit price times the number of permits needed to legitimize remaining emissions) and the second term is control cost.[1]

Permit price would be a function of the behavior of both sources. Since the second source, by assumption, is a price taker, it will seek additional permits until its marginal cost of control is equal to the permit price. This implies $P = 2Q_2$. Furthermore, we know that six units of reduction are needed, so $Q_1 + Q_2 = 6$. Putting these facts together allows us to explain price solely in terms of the first source's behavior.

$$P = 2(6 - Q_1) \qquad (2)$$

Substituting this into (1) yields

$$\min_{Q_1} 2(6 - Q_1)(5 - Q_1) + Q_1^2/2 \qquad (3)$$

Since this equation is expressed solely in terms of the choice variable, (Q_1), it is a simple matter to derive the necessary and sufficient condition for Q_1 to minimize financial burden:

$$4Q_1 - 22 + Q_1 = 0$$
$$Q_1 = 4.4 \qquad (4)$$

1. The total control cost can be found by integrating marginal cost. Since the marginal cost is Q_1, the total cost is $Q_1^2/2$ plus a constant (the fixed cost). In this example the fixed cost is assumed to be zero for simplicity.

The price-setting source would minimize its financial burden by choosing 4.4 units of control, implying (from equation 2) a price of 3.2. This immediately implies that the second source would control 1.6 units (from $2Q_2 = P = 3.2$).

Before examining the nature of this solution, let us be clear about what is going on. Equation (4) expresses the cost-minimizing choice as the one which equates the marginal expenditure on permits, *taking the effect of further purchases on price into account,* to the marginal cost of control. Every additional permit purchased by the first source would drive the price higher, not merely for the additional permits, but for all permits. Therefore to hold price down, the price-setting source must purchase fewer than normal permits, implying that its control costs would be higher than normal. These impressions are confirmed in table 14 where the values of key variables for the two sources are compared for a competitive and noncompetitive auction market.[2]

Several key insights can be gained from this simple numerical example which can serve to focus our inquiry in more realistic situations:

1. Permit prices are lower in the noncompetitive than in the competitive auction market.

2. The price-setting firm controls more emissions than it would if it were merely acting as a price-taker. Because total emissions from all sources would be the same in competitive and noncompetitive markets by the design of the permit system, the price-taking source would have to control fewer emissions in competitive than noncompetitive markets.

3. The noncompetitive auction market allocation of control responsibility is not cost effective; control costs are higher in noncompetitive markets.

4. In terms of percentage change, the impact of noncompetitive behavior on permit prices is quite a bit greater than the impact on total control cost. Though the price-setting source was able to reduce the permit price by 20 percent, control costs rose by only 2 percent.

5. In terms of reduced financial burden, the price-taking source benefits more than the price-setting source. The financial burden for the price-taking source is reduced by 3.3 percent in noncompetitive markets while that for the price-taking source is reduced 16 percent.

2. The derivation of the competitive results are straightforward since the market will be cost effective. This implies $MC_1 = MC_2$, which further implies $Q_1 = 2Q_2$. Using $Q_1 + Q_2 = 6$ yields $Q_1 = 4$ and $Q_2 = 2$.

Table 14.   Competitive and Noncompetitive Auction Markets:
A Numerical Example

Variable	Competitive auction	Noncompetitive auction
Emissions controlled		
Source 1	4	4.4
Source 2	2	1.6
Permits purchased		
Source 1	1	0.6
Source 2	3	3.4
Permit price	4	3.2
Permit expenditures		
Source 1	4	1.92
Source 2	12	10.88
Control costs		
Source 1	8	9.68
Source 2	4	2.56
Total financial burden		
Source 1	12	11.60
Source 2	16	13.44

Source: Calculations by the author based on parameters described in the text.

To those envisioning the price-setting source as inflicting significant harm on other less aggressive sources, these findings may appear surprising. Control costs do rise when one source becomes a price-setter, but no harm is inflicted on other sources. The higher control costs are due to the price-setting firm assuming more of the control responsibility for itself; the greatest benefits are derived by the price-taking, not the price-setting, source.

In many ways the subsidy approach is the mirror image of the auction approach, so it should not be surprising that the presence of a price-setting firm in that setting produces results that are similar. In a subsidy approach, the price-setting firm would try to raise the credit price above its competitive level by holding back on the number of permits sold to the government. This implies that in a comparison of a noncompetitive subsidy allocation with a competitive subsidy allocation, the non-competitive credit price would be higher, the price-setting source's control cost would be higher, the government's expenditures would be higher, and the price-taking source would gain more than the price-setting source.

The theory providing the foundation for this analysis and the numerical example illustrating the application of that theory are also helpful in isolating the factors that determine the degree to which a source can manipulate an auction or subsidy market for its own gain. One such factor is the relative importance of the price-setting source's demand for permits compared with those demands of the competitive, price-taking fringe. When the price-setting source's demand for permits in the neighborhood of the cost-effective allocation comprises only a small proportion of the total demand, manipulating its demand will have little or no influence on the price. Unless the source controls a large proportion of the total demand, its efforts are for nil. Large changes in the demand for a relatively small proportion of the total permits would have only a slight impact on prices.

The shapes of the cost functions for both price-setting and price-taking sources are also important. The slope of the price-setting source's control cost function determines the degree to which it can gain from manipulating the price. Since the price-setting source secures all decreases in price by controlling more emissions (demanding fewer credits), the cost of this strategy to the source pursuing it is determined by the marginal cost function. The higher the marginal costs of control beyond the cost-effective allocation, the less the source is able to gain from a given reduction in demand because the cost of producing the compensating emission reduction is so high.

The cost functions for the competitive fringe are also important. Suppose, for example, that the demand for permits by the competitive fringe was perfectly inelastic (price insensitive). In this case the price-setting firm could act as a monopsonist, lowering price to zero while obtaining all the permits it needed. While the *financial burden* of all sources would be greatly affected by price manipulation in this case, control costs would not. The noncompetitive control costs would be identical to the competitive control costs, since all sources would be controlling exactly the same amounts in both allocations.

The power of any source to lower its financial burden would be diminished when the demand for permits by the competitive fringe was quite price sensitive. In this case, any given price reduction would trigger a larger increase in the demand for permits by these sources, causing a larger increase in the need for the price-setting source to control its own emissions, a costly requirement.

While the directions of these effects are clear from theory, the magnitudes are not. Does the conclusion that total control costs are not much affected by the presence of market power stand up in more realistic settings? Since this conclusion is pivotal in assessing the significance of market imperfections, the answer is of some importance.

Unfortunately, few studies address this issue. One that does (de Lucia, 1974, pp. 124–131) examines the effect of market power on permit prices and control costs on eight sources emitting two different water pollutants (biomass potential and biochemical oxygen demand) into the Mohawk

River.[3] In these two simulations (one for each pollutant), the largest source is assumed to act as a price-setter, with the remaining sources acting as price-takers.[4] In each simulation the designated price-setting source was more than three times larger than the next largest source, accounting for over 46 percent of the uncontrolled total biomass potential and over 45 percent of the biochemical oxygen demand load placed on the river.[5] By arranging its bids, the price-setting source is assumed to manipulate the price so as to minimize its financial burden.

In the simulation of biochemical oxygen demand, the price-setting source has only a negligible effect on price and on the regional cost of controlling all sources. Its attempts to lower the price of permits were effectively thwarted by the cost of the additional control it would have to bear. Its control costs were so large as to prevent it from securing any gain from price manipulation.

In the simulation of biomass potential, the price-setting source does gain a somewhat lower financial burden than it would have if no source acted as a price-setter, but the effect on regional control costs is still negligible. They rise less than one-fifth of 1 percent.

De Lucia also measured the sensitivity of his conclusions by considering a third simulation involving a hypothetical market containing only two polluters—one price-taker and one price-setter—emitting roughly the same uncontrolled levels of waste. In this simulation, permit price was found to be quite sensitive to market structure, falling in the noncompetitive market to less than half its competitive level. The results of this halving of price on control costs and financial burden are depicted in table 15.

When the first source acts as a price-setter rather than a price-taker, total financial burden is reduced substantially for both sources. The noncompetitive financial burden for the two sources together is 19 percent lower than the competitive financial burden. As expected from the numerical example discussed earlier in this chapter, the financial burden of the price-taking source is reduced more, falling some 27 percent, while that of the price-setting firm falls only 14 percent.

The most interesting finding from the point of view of assessing the seriousness of imperfections in auction markets concerns the effect on control costs. Even when one source, controlling roughly half the

3. Biomass potential is a pollution aggregate. It is constructed as a weighted sum of biochemical oxygen demand, total nitrogen, and biologically available phosphorus.

4. To some extent these two experiments are artificial as well since they involve eight municipalities, not a typical situation for air pollution control. Nonetheless, it does involve eight sources and uses more realistic cost functions, which makes it at least somewhat more realistic than our numerical example.

5. de Lucia (1974, p. 86).

Table 15.   Noncompetitive Function Markets and Financial Burden:
The Mohawk River

	Permits purchased	Control cost	Permit cost	Total financial burden
Noncompetitive market				
Source 1 price-setter	217	327,891	42,789	370,680
Source 2 price-taker	283	142,446	55,804	198,250
Total	500	470,337	98,593	568,930
Competitive market				
Source 1 price-taker	245.47	314,769	116,337	431,105
Source 2 price-taker	254.53	152,478	120,635	273,112
Total	500	467,247	236,972	704,217

Source: Adapted by the author from Russell J. de Lucia, *An Evaluation of Marketable Effluent Permit Systems,* Report EPA-600/5-74-030 to the U.S. Environmental Protection Agency (September 1974) table 6-11, p. 129.

market, aggressively manipulates auction market prices in its own interests, control costs rise by less than seven-tenths of 1 percent.

It is useful to place the increases in control cost which do result from market manipulation in the context of the savings to be achieved. Whereas the efficiency losses from market manipulation in these examples are commonly in the neighborhood of 1 percent and do not exceed 2 percent, the typical potential savings from permit markets are substantially higher. This form of market manipulation cannot realistically be used as a reason to avoid auction or subsidy transferable permit markets.

Though the available evidence is very limited, it is remarkably consistent with the numerical example in supporting the notion that control costs are not very sensitive to market manipulation. We have examined a very congenial setting for market manipulation, one involving few sources, low marginal control costs for the price-setting source, and high control costs for the competitive fringe. Despite finding circumstances where prices and total financial burden were dramatically affected, regional control costs were remarkably insensitive to market manipulation in all simulations.

Two other potential fears about price manipulation can be laid to rest. Regardless of the circumstances, price manipulation does not affect air quality because, by design, the auctioned permits hold air quality constant. Furthermore, in either the auction or subsidy approaches, less aggressive price-taking firms are not harmed by price manipulation; they are benefited by lower financial burdens to an even greater extent than is the price-setting source. The only financial loser from this type of market manipulation is the government. It would gain less revenue from a manipulated auction market and would have to make larger payments in a manipulated subsidy market.

For any specific auction or subsidy market which seems particularly vulnerable to price manipulation, special auctions can be designed to counter this power.[6] Known as *incentive-compatible auctions,* they eliminate any incentive for a source to unilaterally use its own bids to control price.

The procedures of an incentive-compatible auction are simple enough. As in a regular auction, each source submits to the auctioneer a demand curve for permits, listing the number of permits desired at each of a number of possible permit prices. The auctioneer sums these bids, chooses the market-clearing price, and awards the permits to those bidding at least as high as the market-clearing price.

So far the procedures are identical to those employed in a conventional auction system. The difference arises in determining the prices to be paid for the acquired permits. In contrast to a single-price auction where all successful bidders would pay the market-clearing price for all awarded permits, the prices paid for n permits acquired in an incentive-compatible auction by any source would be equal to the n highest rejected bids submitted by other sources. Because these rejected bids are by definition lower than the market price (otherwise they would not have been rejected), incentive-compatible auctions imply lower permit expenditures than the more traditional single-price auctions.

This method of price and quantity determination eliminates the price-setting source's incentive to lower the price in an auction by understating its own demand. Although any source understating its demand could acquire fewer permits, it could not lower the price it pays for permits in the absence of collusion with other sources. Because price is determined by the rejected bids of other sources, no source can unilaterally affect the price it pays for permits by artificially raising or lowering its demand. It can only raise its control costs by attempting to influence the process.

Grandfathering

Auction and subsidies are only two forms of permit markets, however, and the emissions trading program is more closely related to a grandfathering approach. The strong conclusion about the lack of sensitivity of regional control costs to price manipulation cannot automatically be generalized to include grandfathered permit schemes.

6. The basic work on the properties of these auctions in an emission trading setting can be found in Lyon (1980, 1982).

Hahn (forthcoming and 1982) has shown that the free distribution initial allocation can change the nature of the market power problem in significant ways. The most important of his results demonstrates that the potential for market power to be exercised is a function of the particular baseline allocation of control responsibility chosen by the control authority. Intuitively, this result is obtained because the ability to exert market power depends on the degree to which the source can affect the price of credits actually traded. The number of credits traded as well as the degree to which the aggressive source can dominate these trades depends on how the pretrade control responsibility is allocated.

Several other conclusions follow from Hahn's analysis of the free distribution case:

1. If the grandfathered baseline control responsibility is cost effective, the existence of one or more price-setting sources would not raise total control costs. Because no trades would take place in this case, there is no opportunity to exert the power.

2. Whenever a single price-setting source receives a baseline control responsibility either exceeding or falling below its cost-effective allocation, total control costs would exceed their minimum. When the baseline control responsibility falls below the cost-effective allocation, the price-setter can exercise power on the selling side and when it is above, the price-setter can exercise power on the buying side of the market.

3. As baseline control responsibility is hypothetically transferred from price-setting to price-taking sources, the price which governs credit trades would rise and the number of credits retained by price-setting firms for their own use would increase.

4. In a grandfathered credit market, the ability of any one source to affect permit prices is a function of its net demand for or net supply of credits (determined by the baseline control responsibility), not the size of the source *per se*.

These results suggest that the flexibility that control authorities have in principle in allocating the financial burden could be substantially less in practice if market power is a potential problem. Deviations from the cost-effective control baseline could cause control costs to exceed their minimum level by opening price-manipulating opportunities. The crucial question is how sensitive control costs are to these deviations. If they turn out to be insensitive, then the control authority's flexibility is not seriously jeopardized.

Two simulation studies have examined this issue. Hahn (forthcoming) has examined the sensitivity of control costs to changes in the initial allocation of control responsibility to one potential price-setting source, using data on sulfur dioxide control in Los Angeles. The results are reproduced as figure 8.

The cost-effective initial allocation of allowable emissions is indicated as Q^*. Control costs are very flat in the neighborhood of that allocation. The one range of initial allocations on the graph which shows a large increase in control costs involves allocating a very large proportion of allowable emissions (or a small proportion of baseline control responsibility) to the price-setting source. In this case, the price-taking sources are all on very steep portions of their marginal cost functions, while the price-setting source is on a relatively flat portion of its marginal cost function. Add the fact that the price-setting firm, by virtue of its high initial allocation of allowable emissions, controls a substantial proportion of the credits available for sale, and it becomes clear why control costs are affected to such a high degree. The price-setting firm has the other sources over a barrel. Should they fail to acquire additional credits, their control costs would be so high as to threaten their continued existence.

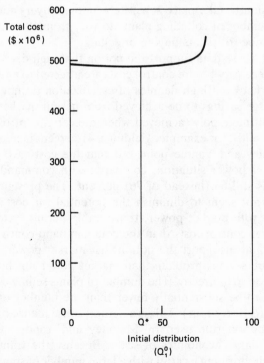

Figure 8. Total annual abatement cost vs. initial permit distribution

Though it appears from figure 8 that monopsony (created by allocating too few permits to the price-setting source) is not a problem, that conclusion may be sensitive to the situation modeled. In other work, Hahn and Noll (1982, pp. 135–137) examine, in the context of sulfate control in Los Angeles, the effect of choosing a control baseline in which one large source is required to control all emissions while other sources control less than their cost-effective amount. This construction creates a situation in which the large source is the only buyer, facing a number of credit sellers. Calculations of the increase in control cost due to this form of market power were performed for a number of cases involving different assumptions about natural gas availability and levels of desired air quality. These calculations showed the losses to be relatively small, ranging from zero to 10 percent, depending upon the case examined.

In another set of published data from the Du Pont Corporation involving some 52 plants and 548 sources of hydrocarbons, Maloney and Yandle (in press) investigated the effects of cartelization of plants on the permit market. Assuming that all sources receive a proportional initial distribution of the permits based on their uncontrolled emissions, they calculate the effects on control costs if plants collude. Their analysis allows collusion to take place separately among buyers and sellers and allows the number of colluding plants to vary from 10 to 90 percent of the total number of plants buying or selling.

In general, these data support the notion that high degrees of cartelization are necessary before control costs are affected to any appreciable degree and that even high degrees of cartelization do not significantly erode the large savings to be achieved from permit markets. At the 90 percent credit monopoly (achieved when the cartel controls 90 percent of all credits sold), for example, yielding a 41 percent increase in control costs, Maloney and Yandle point out that the cost savings from this severe market power situation, compared with command-and-control regulation, is still 66 (instead of 76) percent. The presence of market power does not seem to diminish the potential for cost savings very much. Even with market power, transferable permit systems seem to result in lower control costs than the command-and-control allocation.

These data also support the notion that market power on the seller side is a more serious problem than market power on the buyer side, though they do so indirectly. The number of plants selling credits (eight) is estimated to be substantially fewer than the number of plants purchasing credits (forty-four). This is a natural consequence of the proportional distribution rule used by Maloney and Yandle, which favors sources with large economies of scale. Because the transaction costs associated with forming a cartel with a large number of small sources are significantly greater than those for forming one with a small number of large sources, proportional initial allocation rules make power on the seller side more likely than on the buyer side by creating a situation involving a few plants selling permits to a much larger number of buyers. Other rules could lead to power on the buyer side only if they created a few buyers and a large number of sellers. No realistic published rule evaluated to date reports finding that situation.

Direct data on zero-revenue auctions, the other main form of grand-fathering, are not available. Though in principle the empirical results for normal auctions should apply to zero-revenue auctions as well, these are more complicated institutions and these complications could introduce new unanticipated sources of deviation from the least-cost solution. To provide some initial, admittedly tentative evidence on the seriousness of this problem, Hahn (1983) designed a series of experiments based on techniques drawn from the emerging field of experimental economics.[7] Using students at the California Institute of Technology as subjects, Hahn gave each of eight students a payoff schedule corresponding in shape to one a typical source might face in a zero-revenue auction. The students were then asked to write down the quantity demanded of a fictitious commodity at each price. Hahn then calculated the equilibrium price based on these demands and the resulting payoff to each of the students. They received this payoff in cash.

Three separate simulations were run using a total of twenty-four students. A different initial allocation of permits was used for each of the three simulations.[8] Each simulation was repeated ten times to allow for learning to take place.

The success of the zero-revenue auction is measured by an index which takes on the value 1.0 if the least-cost allocation were achieved and 0.0 if the allocation was no better than the initial allocation of permits. Upon examining this index, Hahn found that when the initial allocation was quite far from the least-cost allocation, the zero-revenue auction did very well. By the tenth iteration of the second and third experiments (involving higher cost initial allocations), the indices were 0.99 and 0.98 respectively. The markets essentially converged to the least-cost allocation. Subsequent experiments by Hahn have yielded similar results.

Based on the theoretical and empirical evidence surveyed, the zero-revenue auction is a very attractive alternative to the approach taken by the emissions trading program, particularly for volatile organic compound trades where location within the airshed is not important. Though

7. For a review of the design, application, and results of these studies, see Plott (1982).

8. From a theoretical point of view, these initial allocations should affect the distribution of payoffs among students but not the size of the total payoff to all students. In a zero-revenue auction, the initial distribution should not affect the potential for market power. The evidence was consistent with this expectation.

it retains the transferability feature that is so important to cost effectiveness and allows the control authority to exercise some control over the distribution of financial burden, it does so while reducing the potential for price manipulation. Since it is totally compatible with the current approach, it could be initiated in those markets where the potential for market power seems particularly likely.

REDUCING COMPETITION

A rather different market power problem arises when sources use emissions trading as a means to reduce the competition they face in product or input markets rather than as a means of reducing permit expenditures. The preconditions for successful predatory action, the consequences of it, and the public policy means of coping with it are unique to this objective.

Causes and Consequences

Sources attempting to use the permit market to reduce the competitive pressures they face in product markets would acquire additional permits or credits in order to deny their use to competitors. How they implement this strategy depends on the nature of the permit market. In an auction market, they would state higher than normal demands. Though this would elevate both the permit price and the financial burden for all sources, the predatory source would expect to gain in the long run by eliminating competitors or by reducing their competitive edge. As this pressure from competition is reduced, the source could raise the prices of its products, recouping any losses it may have incurred in the permit market.

The effects on an auction permit market are obviously rather different, depending on whether the aggressive source seeks to manipulate prices or reduce competition. Whereas the price-setting firm acquires fewer permits than normal, the predatory firm acquires more. Whereas the price and financial burdens of other sources are lower in the company of a price-setting source, they are higher in the company of a predatory source.

Similar results hold for the zero-revenue auction. The only major difference between the normal and zero-revenue auction is that in the normal auction the government makes more money when confronted by a predatory source than a price-setting source. In a zero-revenue auction, the government does not derive any revenue in either case.

The similarities in the effects of these two different types of market power on the permit market are much greater when the baseline control responsibility makes the aggressive source a net seller of permits. When placed on the seller's side of the market, both price-setting and predatory sources seek to offer fewer credits for sale than would be cost effective. Since both strategies raise the price of the credits and, therefore, reduce the financial burden for the aggressive source, both objectives lead to similar consequences in direction, if not in degree.

The Potential

For emissions trading to be used effectively against competitors in the product market, a necessary condition is for these competitors to be in the same permit market. This means that they should be emitting at least one pollutant in common with the predatory source in the same airshed. Product market competition has no relevance if the objective is purely price manipulation.

When product competitors participate in the same permit market, the possibility for reducing competition exists. To turn the possibility into reality, the predatory source would have to be able to impose its will on the market. There are a number of reasons for believing that this would normally be a rare occurrence.

Few permit markets contain a large number of direct competitors. Not only does a typical airshed contain a number of rather different sources, but in many permit markets electric utilities are the largest sources. Being franchised monopolies, these utilities are not driven by the same concerns as competitive firms; they have little to gain from predatory behavior. Even the industrial sources emitting a particular pollutant in a given area rarely have much overlap in product markets.

This suggests that in most product markets the permit market would be a relatively inefficient vehicle for a predatory source to use to inflict harm on competitors. Many of its competitors in the permit market would not be competitors in the product market. Denying permits or credits to those few competitive sources would raise its financial burden, but would provide the predatory source with much in the way of commensurate gain.

In an auction market, if resale is prohibited, the predatory source would in general have to deny permits to all sources to ensure that the targeted competitor was affected.[9] This would be very expensive,

---

9. While it is conceivable that in a particular market the largest competitor might be especially vulnerable, this fortuitous circumstance would presumably be rare.

especially given the evidence in the previous chapter about the size of permit expenditures.

Whether grandfathering enhances or retards market power depends on the rule used to allocate the baseline control responsibility. If the rule were somehow able to allocate all sources of their cost-effective responsibility, no threat is posed to existing source competitors; no target source would be willing to relinquish its permits to the predatory source. A cost-effective allocation of baseline control responsibility allows the existing potential victims to protect themselves without committing large expenditures of funds to that purpose. They could continue operation without being forced to acquire more credits from a potential hostile seller.

The most serious problem would arise when a predatory source was assigned a small baseline control responsibility relative to a cost-effective allocation. The predatory source could pick and choose among those desiring to buy its permits. By denying requests from competitors and honoring those from the other sources, it could harm competitors while reducing its own financial burden.

New sources are more vulnerable, since they have to purchase credits to be allowed to produce anything. As long as sufficient offsets are available from more than one source, this should not be a problem. In general, the new source bias inherent in forcing new, but not existing, sources to purchase offsets to cover any emissions is probably a more serious barrier to entry than the existence of market power.

Even in this relatively congenial situation for the predatory source, it would be difficult to harm competitors. The victim source could frequently purchase credits from sources other than the predatory source. They would be willing to sell since (unlike the victim source) they would be free to replenish their supply from the predatory source (at least until the ultimate destination of the credits becomes clear to the predatory source). Furthermore, it is not clear that this kind of discriminatory treatment of potential buyers by the predatory source would survive close scrutiny either by the control authority or the courts.

While the use of emissions trading by predatory source markets to reduce competition would be extremely rare, it would not be impossible. There are some permit markets where competitors are particularly accessible, such as the Piceance Basin in northwest Colorado.[10] This region contains the site for all currently planned shale oil production in the country. Because the production of shale oil in this region would generate emissions, by cornering the market in credits for that region, one producer could conceivably prevent others from producing.

The ability of a predatory source to reduce competition is not obvious even in this market. Shale oil faces tremendous competition from other

10. See the exchange on this subject in Ryan (1981) and Tietenberg (1981).

sources of oil. Since the facilities producing these other types of oil (including foreign producers) would not be involved in this permit market, they would be immune to the pressure brought by the predatory source. The existence of the other sources of oil would limit the predatory source's ability to raise the price of its oil, the ultimate objective of reducing competition.

In the price-manipulation form of market power, firms from outside the airshed purchasing credits do not pose a threat. Since the purpose of price manipulation is to reduce the price-setting source's cost of permits or credits, firms outside the airshed do not need any of the permits and hence have nothing to gain by manipulating the price (unless they are merely speculating in permits).

Predatory sources outside the airshed can pose a threat. By acquiring permits or credits and denying them to a competitor, they might be able to gain a decisive competitive edge which could subsequently be turned into a larger market share and larger profit.

One way to cope with this problem is to require banked emission credits purchased by industries outside the region to be used within a certain period of time.[11] The ability to purchase would be contingent upon an approved construction plan. If this grace period between the purchase and use of a permit were too short to allow the source to plan effectively, it could secure an option to purchase from any source holding the permit in the bank. These options are common and effective in land transactions. There is no reason to believe they should be less common or less effective in permit markets.

THE EPA EMISSIONS TRADING PROGRAM

Though in principle market power may not be a serious concern, it is true the current EPA emissions trading program makes market power more likely than it would be in a less restrictive program. Elements of that program designed to achieve other objectives have the unfortunate side effect of increasing the vulnerability of emission reduction credit transfers to market power. Although, due to the lack of empirical evidence, it is not possible to quantify the seriousness of this problem, it is possible to use the foregoing analysis to identify potential problem areas and to assess their significance at least qualitatively.

11. One interesting question that arises in this context is what consideration should be given to other organizations, such as environmental groups who wish to acquire permits or credits to retire them. For a discussion of this question, see Russell (1981), Tietenberg (1981), Oppenheimer and Russell (1983), and Howe and Lee (1983).

The Baseline Allocation of Control Responsibility

The baseline allocation of control responsibility in the emissions trading program is the command-and-control allocation. Two aspects of this approach are of particular interest in considering whether this approach fosters or inhibits market power. The fact that it is a grandfathering scheme means that price-manipulation market power is potentially a more serious problem, depending on the initial baseline control allocation, than if an auction had been used. On the other hand, the command-and-control-based distribution rule used in the EPA emissions trading program probably does not create the kind of situation that is conducive to either price manipulation or competition-reducing market power.

Whereas either auction or subsidy markets place all sources on the same side of the market, some sources are buyers and some are sellers in a grandfathered approach. Grandfathering implies fewer participants on each side of the market. Depending on the initial distribution rule, a few sources could comprise a significant proportion of the buyers or the sellers, a condition conducive to both price manipulation and competition reduction.

As we have seen in the above analysis, the distribution rule which creates the most problems allocates a disproportionate share of permits to a few large sources. Because of economies of scale, these sources can sell permits without incurring large increases in control costs. The purchasing sources, facing a large deficit of permits and very high marginal control costs, are vulnerable to price manipulation and to any predatory source seeking to put them out of business.

Fortunately the distribution rule used by EPA is not likely to create this kind of circumstance. Though it is not consistent with the cost-effective allocation (a precondition for *no* market power of either type), it does seem to assign more responsibility (meaning fewer allowable emissions) to large sources. Because it does take economies of scale into account, at least crudely, the baseline allocation of control responsibility chosen by EPA probably has not introduced a significant threat of market power.

In general, grandfathering is beneficial in protecting sources from predators. Since the command-and-control pretrade allocation is generally economically feasible, existing sources would not be forced out of business even if no other source was willing to sell them any credits. Failure to acquire any permits in an auction market would in most cases mean the closure of the plant.

Restrictions on Trading

The potential for predatory behavior or price manipulation is most serious when the price-setting firm faces little or no competition from other buyers or other sellers. Two aspects of the emissions trading program reduce the degree of competition and thereby increase the potential for market power: the treatment of trades among nonproximate sources and the minimum control thresholds.

Chapter 4 discussed some of the rules used by states to govern trades among sources in different locations. Usually these rules are excessively cautious, requiring offset ratios substantially greater than 1.0 for nonproximate sources. In addition to raising compliance costs by reducing trading opportunities, these rules increase the potential for market power by reducing the number of trading sources.

Rules penalizing trades between distant sources favor proximate sources. Indeed, part of the purpose of these rules is to encourage sources to seek emission reduction credits from nearby sources. By so doing, however, they reduce competition in the market and increase the likelihood that proximate sources can exercise market power. Distant emitters which could serve as an alternative source of permits, thereby limiting the proximate source's ability to raise price or to harm a competitor, cannot compete in the face of high offset ratios.[12] By effectively partitioning the market into zones, these rules reduce one of the natural checks on market power.

The minimum control thresholds provide additional opportunities for price manipulation. By setting lower limits on the amount of the control responsibility that can be met with emission reduction credits, they reduce the net demand for permits. Sources which would purchase more permits, in the absence of these regulations, are forced to control more themselves. Fewer permit purchases mean a thinner market on the demand side.

In practice, this is not likely to be a significant problem. By design, the minimum control thresholds do not reduce the *number of purchasers,* merely the average *number of permits* purchased by each source. All new or modified sources must still secure permits to offset any increase in emissions. Because they do not create a circumstance where the number of purchasers is reduced, the minimum treatment standards do not necessarily reduce the competitiveness of the market very much.

Treatment of New Sources and Plant Closures

In nonattainment areas all new sources are required to purchase offsets from existing sources. To the extent these existing sources can exercise market power of either type, the new source bias would be made worse. How serious a problem is this market power likely to be?

In most cases the existing sources will face some control responsibility because they have to meet the RACT standards. Since there are typically

12. Because location is not a factor for trades involving volatile organic compounds, they are not susceptible to this problem.

a lot of existing sources, it is not obvious that any one of them could exert enough of an influence on the market to affect price or the ability of a competitor to secure permits. The existence of other sources ready to sell credits would limit the ability of any single source to exercise market power of either type.

As more sources move into the area and the degree of control on existing sources is tightened, it would become more difficult to free up additional credits. Fewer sources would be willing to sell credits because of the high cost of further control. Those sources which do have credits to sell would be in an advantageous bargaining position. The strength of this bargaining position would depend on the vulnerability of new sources and on the marginal cost of further control by the selling source.

The vulnerability of new sources would depend on the availability of alternative locations. To the extent that locating in this airshed dominates other possible locations, a price-setting existing source would be able to capture some of the location rent associated with this particular desirable location. The price paid to a monopolistic, existing-source seller by a new source would capture some of the value of the location to the new source. As long as there are other equally suitable locations, however, the existing source would not be able to extract much location rent.

The ability of any existing source to manipulate price also depends on its marginal cost of control. In the circumstance we are discussing, the marginal costs of control can be expected to be very high for existing sources. It would be very expensive to make other emission reduction credits available for sale, thus limiting the capacity of any individual source to influence the market.

There is, however, one major exception. When plants close and the emission reductions are banked, the marginal costs associated with creating the emission reduction credits would already have been incurred. Given the limited availability of credits from existing sources, this situation could give a single source command over a substantial number of the available credits. Banked credits for shutdowns do create the possibility for departing firms to extract location rent from unrelated new sources.

The ability of departing firms to command higher prices depends on the availability of other sources of offsets. To the extent that several sources shut down, no one source will be able to unilaterally exercise much influence over price. The most troubling scenario would involve a single source of offsets (generated by a plant closure that has already taken place) facing new sources that are vulnerable (in the sense that alternative locations are much more expensive and their entrance into the market is totally dependent on the acquisition of the permits).

Though in general the approach taken by the EPA in the emissions trading program has made market power more likely than necessary, the degree to which this manipulation could interfere with the basic objectives of the program seems small. The potential cost savings is sufficiently large that even if markets were quite noncompetitive (an unlikely outcome), the emissions trading program would still be more cost effective than the command-and-control allocation.

Should local control authorities become concerned that a unique circumstance in their trading area has created the threat of market power, the control authority could use its eminent domain authority to purchase (not confiscate!) shutdown credits by providing just compensation. Just compensation would be defined in terms of previous emission reduction credit transactions in that area, the rate of inflation, and other relevant factors. The control authority could then sell the credit to some new source at a price which was sufficient to cover its costs (including administration costs). This option would only be exercised when the only offsets for sale were the shutdown credits and where the market power threat seemed particularly high. Because it would presumably be exercised only when a willing buyer was available, the cash flow implications for the control authority would be minimal.

SUMMARY

• Two rather distinct types of market power are possible in transferable permit markets. The first arises when a price-setting source or a collusive coalition of sources seeks to manipulate the price of permits for their own financial gain. The second stems from the desire of one predatory source or a collusive coalition of sources to use the permit market as a vehicle for reducing the competition they face in product or factor markets.

• In permit auction (subsidy) markets, price-setting sources purchase (sell) fewer permits than is cost effective. Control costs are higher than the least-cost allocation, but the financial burdens borne by all sources (not merely the price-setting source) are lower. Air quality is not adversely affected, but revenues received by (paid by) the control authority are reduced (increased).

• The available evidence suggests that price manipulation in permit auctions under the worst circumstances can have a large influence on price, but a relatively small influence on control costs. In the available empirical studies, control costs typically rise no more than 1 percent above the least-cost solution. Compared with the potential savings estimated in chapter 3, this rise is very small. The factors influencing the ability to manipulate demand include the relative size of the price-setting source's demand in relation to the demands of other sources, and marginal control costs for both the price-setting and price-taking firms.

• In cases where price manipulation in an auction (or subsidy) market could be a problem, a special type of incentive-compatible auction can be used to remove the incentive for price manipulation, at least for noncolluding firms.

• When a grandfathering approach is used instead of an auction or a subsidy, the degree to which prices can be manipulated depends on the rule used to distribute the baseline control responsibility. If somehow the responsibility were distributed such that each source received its cost-effective allocation, price-setting sources would gain no market power and control costs would remain at the least-cost level. As the baseline control responsibility is transferred from price-setting to price-taking firms, the equilibrium price rises, as does the equilibrium number of credits held by the price-setting firm.

• A necessary condition for a predatory source to use its power in the permit market to reduce product-market competition is for the competitors to be emitting the same pollutant in the same airshed. While it is relatively rare for many direct competitors to participate in the same permit market, it can occur. One example is the Piceance Basin in Colorado where all permit purchasers intend to use their permits to legitimize emissions created during the production of shale oil. Even in this extreme and unusual case, predatory behavior is not likely to succeed because these shale oil producers all face direct competition from producers of different types of oil. Since these producers are not in the permit market, the ability of a predatory source to raise its product price to recoup its losses from putting a competitor out of business is accordingly limited.

• In auction markets, whereas a price-setting source reduces its financial burden and those of all other sources in the market, a predatory source raises its financial burden and those of the other sources in the market. Because the permit expenditures are so large and the number of noncompetitors in the typical market are so numerous, this type of market power is unlikely to have much effect in auction or subsidy markets.

• Whether grandfathering enhances or discourages predatory behavior depends on the distribution rule. The larger the number of permits granted to the predatory source above its cost-effective allocation, the more serious the problem. This circumstance allows the source to target the harm, reducing competition without seriously increasing its financial burden.

• In general, the baseline allocation of control responsibility implicit in the emissions trading program, while certainly not equal to the allocation minimizing market power, is sufficiently close, given the lack of sensitivity of control costs to price manipulation, that market power of either type is not normally likely to be a significant problem. The apparent neglect of market power issues in the EPA emissions trading program has probably not had much detrimental impact on its cost-saving potential.

• That is not to say that the program has been neutral. Two aspects of the current program do increase the threat of market power: the rules governing trades between nonproximate sources and the minimum treatment standards. The current trading rules unnecessarily penalize distant sources, reducing the effective size of the market and thereby increasing the possibilities for price manipulation or harming competitors by proximate sources. The minimum control thresholds reduce the demand for credits, making for a thinner market.

• Market power can increase the new source bias, since permit prices can be affected even when control costs are not. The vulnerability of new sources to either type of market power would depend on the availability of alternative production locations and the availability of multiple suppliers of offsets.

• Shutdown credits offer an unusual opportunity for sellers to exercise market power when no other offsets are available. Because this is a selective, not a universal, problem with shutdown credits, a contingent remedy is appropriate. When shutdown credits are being used to manipulate prices, the control authority can use its eminent domain power to purchase shutdown credits by providing just compensation. This is preferable to the current approach (confiscation without compensation) because it preserves appropriate incentives for sources to generate shutdown credits, while preventing both the exercise of market power and any resulting increase in the new source bias.

REFERENCES

Dales, John H. 1968. *Pollution, Property and Prices* (Toronto: University of Toronto Press).
de Lucia, Russell J. 1974. *An Evaluation of Marketable Effluent Permit Systems,* Report no. EPA-600/5-74-030 to the U.S. Environmental Protection Agency (September).

Hahn, Robert W. 1982. "Market Power and Transferable Property Rights" in Glen R. Cass, Robert W. Hahn, Roger G. Noll, William P. Rogerson, George Fox, and Asha Paragjape, eds., *Implementing Tradable Permits for Sulfur Oxides Emissions: A Case Study in the South Coast Air Basin* vol. 3, Appendices (Pasadena, Calif., California Institute of Technology) pp. C45–C70.

———. 1983. "Designing Markets in Transferable Property Rights: A Practitioner's Guide," in Erhard F. Joeres and Martin H. David, eds., *Buying a Better Environment: Cost-Effective Regulation Through Permit Trading* (Madison, Wis., University of Wisconsin Press) pp. 83–97.

———. "Market Power and Transferable Property Rights," *Quarterly Journal of Economics,* forthcoming.

———, and Roger G. Noll. 1982. "Designing a Market for Tradable Emissions Permits," in Wesley A. Magat, ed., *Reform of Environmental Regulation* (Cambridge, Mass., Ballinger).

Howe, Charles W., and Dwight R. Lee. 1983. "Organizing the Receptor Side of Pollution Rights Markets," *Australian Economic Papers* (December) pp. 280–289.

Lyon, Randolph M. 1980. "Auctions and Alternative Procedures for Public Allocation: With Applications to the Distribution of Pollution Rights," unpublished Ph.D. dissertation (Urbana, Ill., University of Illinois).

———. 1982. "Auctions and Alternative Procedures for Allocating Pollution Rights," *Land Economics* vol. 58, no. 1 (February) pp. 16–32.

Maloney, Michael T., and Bruce Yandle. "Estimation of the Cost of Air Pollution Regulation," *Journal of Environmental Economics and Management,* in press.

Oppenheimer, Joe A., and Clifford Russell. 1983. "A Tempest in a Teapot: The Analysis and Evaluation of Environmental Groups Trading in Markets for Pollution Permits," in Erhard F. Joeres and Martin H. David, eds., *Buying a Better Environment: Cost-Effective Regulation Through Permit Trading* (Madison, Wis., University of Wisconsin Press).

Plott, Charles R. 1982. "Industrial Organization Theory and Experimental Economics," *Journal of Economic Literature* vol. 20, no. 4 (December) pp. 1485–1527.

Russell, Clifford S. 1981. "Controlled Trading of Pollution Permits," *Environmental Science and Technology* vol. 15, no. 1 (January) pp. 24–28.

Ryan, Donald R. 1981. "Transferable Discharge Permits and the Control of Stationary Source Air Pollution: A Survey and Synthesis: Comment," *Land Economics* vol. 57, no. 4 (November) pp. 639–641.

Tietenberg, Thomas H. 1981. "Transferable Discharge Permits and the Control of Stationary Source Air Pollution: A Survey and Synthesis: Reply," *Land Economics* vol. 57, no. 4 (November) pp. 642–644.

7 / The Temporal Dimension

The way emission rates vary over time is an important factor in the design of any air pollution control strategy, including emissions trading. This chapter examines how the emissions trading program handles the temporal dimension. Of particular concern are the fluctuations in pollutant concentrations that occur from one period to the next. How are fluctuations in ambient concentration integrated into the goals pursued by the emissions trading policy and the procedures used to achieve those goals? Are the procedures cost effective? Can the temporal component of the current policy be improved?

THE PROBABILISTIC NATURE OF POLLUTANT
CONCENTRATIONS

Sources of Concentration Variation

Pollutant concentrations at specific receptor locations are monitored by taking samples of the air at frequent, regular intervals. Typical plots of concentration frequencies recorded from those samples show the ambient concentration to be distributed as a log-normal random variable.[1] The log-normal distribution is nonsymmetric, with the modal (most frequent) concentration lower than the mean (average). One example of a distribution fitting this description is given as figure 9.

1. See, for example, Larsen (1971).

149

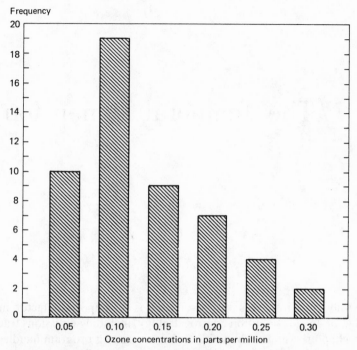

Figure 9. Highest hourly average ozone concentrations in parts per million, Anaheim, California, 1979.

Variation in emission rates is one source of the variation in concentrations. Sources do not emit at constant rates. Some emission rates show a striking seasonal or daily pattern. Space heating or air conditioning emissions are seasonal, while mobile source emissions increase significantly during morning or evening rush hours. Others caused by breakdowns or accidents may be distinctly random.

Variation in meteorological conditions is a second source of concentration variation. Wind speed and direction have both random and cyclical components, including a definite seasonal pattern, particularly for areas near very large bodies of water. Since the water and the land heat and cool at different rates, local onshore or offshore breezes can be created by this temperature differential.

From the point of view of pollution control, thermal inversions are the most adverse weather conditions. They occur when a temperature inversion distorts the normally smooth upward flow of warm air. Robbing the atmosphere of its normal ability to disperse and dilute the pollutants, thermal inversions trap emissions in a small volume of air, creating very high concentration levels.

Concentration variations have important consequences for policy. Since the Clean Air Act bases the primary ambient standards on health, the effects of these concentrations on health must be ascertained. Are short-term, high concentration exposures harmful or is human health more sensitive to cumulative exposure over some longer period of time? How should policy be structured to reduce short-term or long-term exposures?

Defining the Ambient Standards

For each pollutant, the form of the ambient standard has been chosen to mitigate the known or suspected damage caused by that particular pollutant. If short-term, high-dose exposures are found to be dangerous, the standard has been stated in terms of maximum short-term exposures. Short-term exposures are monitored using concentration levels averaged over periods as short as an hour or as long as 24 hours. If the damage seems to be closely related to longer term exposures, then the standard is stated in terms of an annual average.

Several pollutants have multiple ambient standards. (The standards are listed in table 1, chapter 1.) Sulfur oxides and particulates both have an annual average standard and a 24-hour standard. Carbon monoxide has an 8-hour standard and a 1-hour standard. When multiple standards are established, the states must comply with all of them.

The form of the standard makes a difference in defining what strategies can be employed to meet it. Ambient standards based purely on annual averages can be met without worrying about the timing of the emissions. Emission reductions reduce the average whenever they occur. The fact that concentration levels exhibit a series of peaks and troughs over time is of little consequence as long as the standard is based solely on an annual average.

When the ambient standard is based on a short-term average, however, the timing and the quantity of emissions are both important. Since the objective is to reduce the highest concentrations, one strategy is to shift emissions away from the peak concentration periods. Although this strategy would have a large impact on reducing the highest concentrations, it would not necessarily have any impact on the annual average. Indeed, it is conceivable that strategies which are successful in meeting the maximum short-term average standard could leave the annual average concentration unaffected or could even increase it, providing that off-peak emissions increased by a large enough amount.

Temporal strategies (those controlling the *timing* as well as the *level* of emissions) become more attractive as the averaging period is reduced. Since shorter averaging times imply less opportunity to average peak with off-peak periods, reducing emissions during a relatively short period is especially important. To the extent these standards are picking up concentration "spikes"—short duration, high concentration peaks—shifting emissions to other periods becomes easier, since the change in emission timing involved in any shift of emissions from a peak to an off-peak concentration period is rather small.

Because the major component of the variation in observed concentration is regular and, therefore, predictable, temporal strategies are tractable. Peak concentrations for certain pollutants always occur during the same season or even the same time of day, allowing their occurrence to be anticipated and controlled. Peak concentrations can be associated with specific times.

The point is perhaps most easily illustrated by ozone, the pollutant most frequently responsible for air quality control regions receiving a nonattainment designation. The ambient standard for ozone is based on a 1-hour averaging time. One-hour ozone concentrations typically show pronounced diurnal (daily) and seasonal peaks.[2] The entire northeastern United States, from Washington, D.C., to the Canadian border including states as far west as Ohio, is designated as a nonattainment area for ozone, but in this region the exceedances occur only during the summer. The prevalence of ideal conditions for ozone formation (warm sunny days) during that period is responsible. These pronounced daily and seasonal patterns, coupled with a standard defined in terms of a short averaging time, make ozone a prime candidate for controlling the timing, as well as the level, of emissions.

COST-EFFECTIVE TEMPORAL CONTROL

Controlling the timing as well as the flow of goods and services is a familiar activity in public policy. Peak-hour pricing is increasingly required of electric utilities by public utility commissions, not only as a better way of utilizing the existing generating capacity, but also as a way of lowering the demand for new capacity. By charging peak-hour use at a higher rate, these commissions are simultaneously forcing those who create the need for new capacity to bear the cost of building it and to consider switching some of their peak use to off-peak periods as a viable alternative to new construction.

Telephone service and mass transit pricing provide two more familiar examples. Higher rates are charged for calls placed during the peak

2. Ozone concentrations by time of day and month are graphed for particular urban areas in U.S. Department of Health, Education and Welfare (1970, pp. 3-6 thru 3-16).

periods both to finance the capacity expansion needed to meet growth in calls during the peak period and to encourage off-peak use when possible. Higher peak-hour transit fares, common in cities such as Washington, D.C., are designed to accomplish a similar balance between peak and off-peak transit travel.

Although these familiar examples make it clear that temporal control is not a particularly novel concept, they all involve charging higher prices in peak periods. Since the emissions trading program is a quantity-based approach, not a price-based approach, the control authority regulates emissions, not prices. How could cost-effective temporal control be exercised in a quantity-based system?

In thinking about incorporating the timing of emissions into a transferable permit approach, it is important to distinguish two different types of temporal control. The first involves the rather regular, and therefore anticipated, seasonal or daily fluctuations in concentrations. We shall refer to this type of control as *periodic control* to convey the regularity of these conditions. The key aspect of periodic control is that the timing of these conditions can be anticipated. Because the conditions can be identified in advance, the means of controlling emissions during those periods can also be identified in advance.

The second form of control, known as *episode control,* involves irregular concentration peaks which, to the extent they can be anticipated at all, can be anticipated only very shortly before they occur. The timing of these peaks depends on the initiation of thermal inversions or other adverse meteorological circumstances which occur randomly. The necessity for additional control cannot be identified more than a day or so in advance.

Periodic Control

Perhaps the easiest way to describe cost-effective temporal control is to compare it with constant control. Constant control defines the control responsibility in terms of a temporally invariant allowable concentration level and a temporally invariant mix of source responsibilities for meeting that level. In contrast, both the allowable concentration level and the mix of source responsibilities vary over time with cost-effective temporal control.

The recorded pollution at any monitor is composed of two elements: background and controllable pollution. Background pollution results from sources not under the control of the local control authority either because they are unregulated or because they are located outside of the control jurisdiction. Controlled sources must make up the difference between the observed reading and the ambient standard. As the background pollution levels change (such as when ozone transported from other regions increases), the remaining allowable concentration changes as well.

The mix of source controls changes over time for a variety of reasons. Costs of control for each source vary across seasons or even over the 24-hour daily cycle. For nonuniformly mixed pollutants, the transfer coefficients may change as a result of shifts in the prevailing meteorology or different monitors may be recording exceedances in different seasons, necessitating changes in the location of emission control over time.

How would a temporally cost-effective permit system take these considerations into account? For uniformly mixed pollutants, such as volatile organic compounds, location is not important, so transfer coefficients play no role. The first step in controlling these pollutants would be to designate the number of permit periods. This could be as simple as two periods (corresponding to a peak concentration and an off-peak concentration period) or as complicated as necessary. Using the chemical reactivities, meteorological conditions, and background pollution levels unique to each period, allowable emissions would be scaled to meet the standard, with individual permits designed to conform with that allowable emissions level.

Consider how this system would apply to the control of volatile organic compounds, a prime contributor to the formation of ozone. Because ozone formation depends on temperature and sunlight, it is not a problem in the North during the winter months. The warm months could be designated as the peak period and the cold months as off-peak. The boundaries between the peak and off-peak periods would be defined in terms of temperature or sunlight and would vary from region to region. More baseline control responsibility would be assigned in the peak period. Trades reducing emissions during the peak period would be especially encouraged.

For nonuniformly mixed assimilative pollutants, the permits would be defined in terms of allowed concentrations (rather than emissions). The number of permits in each period would be defined in terms of allowed concentrations in each period. Fewer permits would be allocated when allowed concentrations were small. Whereas uniformly mixed pollutants could be traded on a ton-for-ton basis, trades among sources involving nonuniformly mixed assimilative pollutants would have to use the transfer coefficients pertaining to the particular period of the trade. Trades involving concentration reductions during the peak period would have to use transfer coefficients computed especially for that period.

The essence of this system is to tailor the degree of control to the need for it. Sources would be confronted with the need to control emissions especially vigorously during those periods where the most control was needed to meet the standards. Because of their relative scarcity, the permits for the peak period would be higher priced. Sources would have an incentive to add extra control during those periods, to switch some of their emissions to off-peak periods, or to control more for all periods, selling some of the surplus off-peak control.

Constant control, the alternative, ignores the temporal dimension and designs emission standards with sufficient stringency that the concentrations fall within the standards even under the most adverse meteorological circumstances. In terms of figure 9, a constant control strategy would shift the entire distribution of concentrations toward the origin whereas a periodic control strategy would reduce the right-hand tail of the distribution.

Constant control is an excessively costly means of reaching a short-term standard for two main reasons: (1) it requires larger emission reductions than necessary to meet the ambient standards and (2) it does not take into account which sources can control most cheaply *at the time the control is needed*. Larger emission reductions are required by a constant control strategy because all sources are required to undertake a degree of control that is sufficiently stringent that the ambient standard is met under the most adverse conditions. This "worst case" approach requires more control than necessary in less adverse circumstances.

Because it allocates the control responsibility among sources so as to minimize the cost of constant control rather than the cost of periodic control, a constant control strategy fails to identify and to take advantage of sources which can exercise control most cheaply during those periods when the largest amount of control is needed. There may well be sources which could be easily induced to exercise more control during the peak period with a higher peak permit price. Without allowing the permit price to reflect the time variation in the difficulty of control (as would be the case with constant control), sources which could control most easily during that period would never be identified. As a result, they control too little during that period while other sources control too much.

Unfortunately there are no air pollution studies that shed any light on the magnitude of the cost savings due to the use of periodic rather than constant control. Fortunately there have been some water pollution studies. Though the circumstances are sufficiently different in air and water pollution control to preclude using the results from one to draw firm conclusions about the other, the results are suggestive.

One case study by Yaron (1979) involved two reaches of the DuPage River in Illinois that were being polluted by two treatment plants and four industrial plants. The model contained two seasons and simulated the achievement of a DO standard by both constant and periodic control. The results suggest that periodic control is significantly cheaper. The total variable cost was $400,405 per year for the constant control, but $137,046 less for the periodic control, a savings of 34 percent.

A second study by O'Neil (1983) is particularly helpful because of its detailed treatment of capital costs and source location in a multiple receptor framework. Simulating BOD emissions into the Fox River in Northern Wisconsin for each of three periods, he compared a periodic policy with a "worst case" temporally invariant rule. The Wisconsin Department of Natural Resources defined the worst case adverse flow and temperature conditions, and a constant matrix of transfer coefficients derived from these conditions was used to govern trades in all three periods. The periodic case involved different transfer coefficients in each of the three periods.

Two principal results were obtained. First, the "worst case" rule based on flow and temperature conditions failed to prevent the standard from being violated at one of the receptors during one of the off-peak periods. This result underlines an additional problem with constant controls—it is difficult to select in advance a single transfer coefficient matrix and a single number of permits that will achieve the standards in all periods.[3] By using transfer coefficients and allowable emissions tailored to each period, the periodic policy reduces this problem.

The actual concentration in any period is a function of the background concentration plus a weighted average of the emissions from all sources, where the weights are the prevailing transfer coefficients. A constant policy uses only one set of transfer coefficients for the entire year, which would be accurate in only one period (the designated worst case period). Therefore the use of a single constant matrix opens the distinct possibility that the standards could be violated in the other periods. In the case study by O'Neil that is precisely what happened. Constant policies increase the risk of violating the ambient standards.

Although for most situations there would exist a true worst case that could eliminate the problem, in general the control authority cannot define that worst case in advance. It depends not only on the allowable concentration increases and the transfer coefficients (both of which can be determined in advance) but also on the distribution of emission loadings among sources for each period (which cannot be determined in advance). Therefore the control authority has only two undesirable choices with a constant policy—it can either include a significant margin of safety (extra control), which is expensive, or it can run the risk of violating the ambient standards, which over the long run could be even more expensive.

---

3. This result can be most easily seen in O'Neil (1980, pp. 129 and 141). This is a more complete description of the same model employed in O'Neil (1983).

The O'Neil study also found that the constant "worst case" scenario resulted in control costs that were approximately the same as those for the periodic policy. This result is due to the type of abatement modeled in this study in which variable costs were quite small in comparison with fixed costs. Since only variable costs can be saved, the dominance of fixed costs leads to small estimated savings. Though their costs are not strictly comparable because the periodic policy met the standards while the constant policy did not, this result does point up the importance of the balance between fixed and variable costs.

In practice, existing firms with installed equipment would find the cost savings from temporal control more difficult to capture than would new sources. Even with a periodic control policy, the capacity of any capital control equipment would have to be tailored to the period representing the most severe emission reduction requirements for that source, leaving excess capacity in the other periods. Because this capacity has to be financed whether or not it is fully used, nothing is saved when installed capacity is underutilized. In contrast, operating costs are typically variable and, therefore, can be saved in periods of lower control. Because sources with high variable and low fixed control costs would be able to take more advantage of a periodic policy, capturing more cost savings, the ability to save costs with periodic policies depends on the mix of control technologies. Periodic policies encourage the development of more flexible technologies and accommodate their flexibility.

The cost of a constant control strategy seems to be quite sensitive to the stringency of the short-term standard. The more stringent the standard, the greater the cost of constant control. Investigating the sensitivity of the cost of controlling nitrogen dioxide in Chicago, Anderson, Reid, and Seskin (1979, pp. 5-28 to 5-34) found that using a constant control strategy, a 500 g/m^3 1-hour standard could be attained at an annual cost of $1 million while $24 million would be required to reach a 250 g/m^3 1-hour standard. Though they did not analyze how much this cost would be reduced if periodic instead of constant control were used, they did find that the cost of constant control rises very rapidly as the standard is made more stringent. Because meeting more stringent standards with constant controls creates larger excess control capacities during the off-peak periods, the potential for large savings from periodic controls is enhanced as long as flexible technologies are available.

One possible concern with periodic controls is that their use could lead to larger emission loadings on the environment. By shifting emissions from peak concentration periods to other periods, periodic controls may result in more emissions. Can a workable plan be offered in the face of this objection?

The emission loading issue can be dealt with through the judicious use of long- and short-term ambient standards. All emissions enter into the calculation of the annual average, so it provides a convenient check on emission loadings. As long as the ambient standard is defined in terms of a short-term and an annual average, the annual average would control emission loadings while the short-term average would be used to ensure that health is protected from high short-term exposure. For those pollutants having standards defined only in terms of a short-term standard, an annual average could be added as needed. For those pollutants having annual averages, the levels could be made more stringent to reflect their new purpose if that is necessary.

Care would have to be taken for pollutants that can be transported long distances to ensure that emissions from tall stacks would not escape this accounting device. Otherwise, unwanted higher emission loadings could occur with periodic approaches. This can be handled by limiting the credit granted to stack height in modeling the effect of emissions on the annual average.

Because constant control can be such an expensive way to control short-term exposure, it diminishes EPA's enthusiasm for setting new short-term standards or for increasing the stringency of old ones. The short-term NO_2 standard is a case in point. Despite some evidence of adverse health effects resulting from short-term exposure to NO_2, EPA has been very reluctant to approve a short-term standard.[4] Given the uncertainty lying behind the health effect estimates, the costs of meeting the standard become an implicit factor in the decision.

To the extent the NO_2 decision is typical, the prohibition against periodic control has had exactly the opposite effect from what was intended. While seeking to maximize emission reductions, the policy has in practice resulted in higher emission loadings by increasing the reluctance of EPA to impose new standards in the face of very high control costs. Though prohibited by the Clean Air Act from explicitly considering costs, there is little doubt that costs are playing an implicit role in these decisions.

Unreasonably high control costs also open the door to variances. Faced with crippling costs, sources seek to reduce their burden through the courts. Their chances for success are enhanced when the costs seem totally out of proportion with the benefits. Approved variances result in higher emission loadings. Because periodic control costs less, the demand for variances and the chances of variances being approved are diminished. While in principle emission loadings could be higher with periodic policy, in practice they could well be lower.

---

4. One description of the ongoing debate between EPA and OMB over a proposed short-term NO_2 standard can be found in *Inside EPA Weekly Report* (December 23, 1983) p. 1. Similar concerns have arisen with respect to the TSP standard. See *Inside EPA Weekly Report* (December 9, 1983) pp. 1, 7–9.

Episode Control

To the extent that thermal inversions or other devastating meteorological circumstances occur regularly at the same times during the year, in principle they could be handled by a periodic policy. In practice that is not possible because the occurrence of these conditions is not regular and the degree of control required is so severe. This high degree of control should be imposed only when needed despite the fact that the periods of need cannot be identified more than a day or so in advance. Episodes deserve separate treatment.

In order to be cost effective, episode control policies need to identify in advance those sources that can cut back relatively cheaply on very short notice. Meteorologists can give at least one day advance notice and sometimes more as to when the episodes could occur. The basic problem with identifying the amount each source should reduce emissions during an episode is that every source has an incentive to argue for the smallest control responsibility possible. Because they bear the cost of further control, they can hold costs down by avoiding responsibility.

An episode permit system offers one way to resolve this problem.[5] It could be a relatively simple program whenever location could safely be ignored. Unless the episodes at some particular location tend to be triggered by one particular source, the primary object is to reduce total emission loadings in a region. This is the objective most likely to support an episode permit system.

The episode permit system works by assigning priority numbers to emission permits. These can either be assigned continuously for each ton of emissions or, as is probably more practical, assigned to a few priority categories. The higher the priority of the right, the lower the probability that its use would be prohibited during an episode. The highest priority permits would allow uninterrupted continuous emissions. In terms of two priority categories, for example, during a priority one period, only priority one permits could be used. During a priority two period, either priority one or priority two permits could be used.

The control authority would be responsible for defining the conditions which would trigger a priority one, or priority two alert as well as announcing via a prearranged communication channel when these alerts were in effect. It would also be responsible for defining the number of permits (i.e., the amount of allowed emission) during each type of alert and enforcing the ultimate allocation.

The sources would then secure the appropriate number of permits of each priority designation by trading among themselves. Since priority one alert designation permits would allow uninterrupted continuous

5. My thinking on this subject was significantly influenced by Howe and Lee (1983).

emissions, they would command the highest price. Many sources could be expected to purchase some of each category of permits, reflecting their costs of short-notice reaction. They would cut emissions back during each type of alert until the marginal cost of control was equal to the permit price for that alert.

With this system in place, the control authority would announce the alert number and all sources would reduce emissions to the predetermined levels established for that alert. Because of the different prices of permits, sources would voluntarily sort themselves out by their costs of short-notice control. Those who could control relatively cheaply on short notice (say by switching fuel or using afterburners for volatile organic compounds) would save money by not having to purchase the priority one permits. Others who could not reduce emissions as cheaply would purchase the permits and continue emitting during the episode.

The episode permit system not only provides incentives for existing sources to make short-notice reductions as cheaply as possible, it also provides incentives to develop and to adopt control technologies which allow this kind of flexible response. By adopting more flexible control systems, sources could lower their expenditure on episode permits. By allowing sources to save money as they adopt flexible-control technologies, control authorities give the manufacturers of these technologies an edge over competitors who are producing less flexible control technologies. The cost savings from an episode control policy using transferable permits could be expected to rise over time.

Only one published empirical study attempts to establish the magnitude of savings possible from using an episode control policy tailored to the circumstances rather than a constant control policy that is sufficiently stringent to preclude the episodes under any circumstances. Teller (1970) examined the cost of controlling sulfur dioxide in Nashville, Tennessee using fuel substitution for constant and episode control strategies. Although both strategies were designed to ensure attainment of the same ambient standard, the constant control strategy was found to be five times more expensive than the episode control strategy.

THE EPA EMISSIONS TRADING PROGRAM

In no area of permit design is the current practice farther from the dictates of cost effectiveness than it is in the temporal dimension. The philosophy behind the Clean Air Act dictates constant, continuous control supplemented only by a rather weak program for episode control lying in the wings. The degree of constant control has in general been so stringent that the episode controls have in recent years rarely, if ever, been invoked. As the Teller (1970) study suggests, this is a very expensive approach.

Periodic Control

The limits imposed by the Clean Air Act on periodic controls could not be clearer or less congenial. The definitions section of the act states:

> The terms "emission limitation" and "emission standard" mean a requirement established by the State or Administrator which limits the quantity, rate, or concentration of emissions of air pollutants on a continuous basis, including any requirement relating to the operation or maintenance of a source to assure continuous emission reduction. (42 USCS 7602(k))

In conformance with these definitions, the new source performance standards are defined in terms of continuous emission control, as are BACT and LAER.

Another section of the act removes any lingering doubt about its view of periodic control:

> The degree of emission limitation required for control of any air pollutant under an applicable implementation plan under this title shall not be affected in any manner by—
> (1) ...
> (2) any other dispersion technique. (42 U.S.C. 7424(a)) ... For the purpose of this section, the term "dispersion technique" includes any intermittent or supplemental control of air pollutants varying with atmospheric conditions. (42 U.S.C. 7424(b))

In general, the courts have supported a strict interpretation of these sections. In two cases, *Big Rivers Electric Corp.* v. *Environmental Protection Agency*[6] and *Kennecott Copper Corporation* v. *Train,*[7] the court supported EPA's decision to disallow periodic control *even when these plans reduced total emissions as well as improving air quality in the worst periods.*

A case decided after the 1977 Amendments to the Clean Air Act were passed made it clear that with the current statute periodic control could not be used even as a temporary measure until permanent controls were installed.[8] Dow Chemical had formulated a plan whereby sulfur dioxide emissions from one of its plants would be permanently controlled in five years. Because the plan could not reasonably be implemented earlier,

6. 523 F. 2d 16 (1975).
7. 526 F. 2d 1149 (1975).
8. 635 F. 2d 559 (1980).

however, Dow sought to use periodic controls to ensure that the ambient standards would be met in the interim. The court ruled against Dow and against periodic controls.[9]

These legal precedents eliminate any possibility of periodic control, not only in the command-and-control system, but in the emissions trading program as well. No shifts in emissions among periods are permitted, even when such trades could result in substantially better air quality during the peak periods.

Though the degree to which compliance costs are raised by a reliance on constant control cannot be ascertained with any certainty given the paucity of studies on this issue, the prohibition against periodic control lowers the potential cost savings. The degree of emission control should be tailored to the need.

Prohibiting periodic control increases the likelihood that the ambient standards will be violated. Because in a constant control approach it is difficult to pick a worst-case scenario in advance, mistakes would jeopardize attainment. There is no guarantee that the selected control, though severe, would be stringent enough to meet the standards in all periods. Excessive expense does not guarantee enhanced compliance.

In some parts of the country where the feasible continuous emission reductions are not sufficient to meet the standards, some additional means to meet the control must be devised. Since the obvious means, periodic control, has been ruled out, these areas may be sacrificing the only possible means of reaching attainment. If the resulting concentrations are unhealthy, this is a high cost to pay.

By ignoring the timing of the reductions, emissions trades could actually make things worse. For example, a volatile organic compound trade from a source emitting in the winter to one emitting in the summer would increase the ozone concentration during the peak period, but would be allowed under the current regulations in many states. Some states, such as New Jersey, have taken steps to prohibit these specific trades, but no general rules have emerged. At present, even among those states where ad hoc rules prevent some trades that could make air quality worse, no state has established rules encouraging trades which would improve air quality during the peak period.

This is an important and practical point. Ozone, for example, is responsible for more nonattainment designations than any other pollutant. It would be relatively easy to set up a system which differentiated volatile organic compound emission reductions occurring during warm months from those occurring during cold months. Trades reducing peak period emissions would be encouraged and trades increasing peak period emis-

9. This case is discussed further in DeBois (1981).

sions would be discouraged. Current rules neither adequately encourage peak reductions nor adequately discourage peak increases.

The point is particularly significant when applied to the development and adoption of new technologies. Society could benefit greatly from having control technologies that could respond to changing conditions, but the current approach does not reward flexibility. Periodic controls are essential if that flexibility is to be forthcoming.

Episode Control

The current command-and-control system includes an episode control program, but it is entirely separate from the emissions trading program. EPA has realized that high concentrations which are especially dangerous to health could occasionally arise in spite of the high degree of constant control called for in the regulations and that these high concentrations may be sufficiently worrisome to call forth extraordinary remedies. The episode control program is their response to this need.

The current system of episode control is implemented through the SIP system.[10] States are required to include a section on episode control in their SIP for all portions of their air quality control regions designated as priority I regions and some regions designated as priority II regions. The priority designations are based on measured air quality levels, with the regions with the worst air quality being priority I.

The episode regulations provide a set of minimum air quality levels at which the special control procedures must be triggered. These air quality triggers involve significantly worse air quality than the primary and secondary standards. States are required to establish their own criteria for two or more stages (the EPA sample regulations suggest three stages: an alert, a warning, and an emergency) which are at least as stringent as the minimum.

In addition to these thresholds, the plans must contain procedures for forecasting when these thresholds would be crossed, for announcing the episode levels as they occur, and for enforcing emissions reductions during these periods. The regulations suggest specific means by which these reductions might be achieved, including fuel substitution, curtailing or deferring production, and the prohibition of open burning, among others. It is up to each state to identify those most relevant to its special circumstance and to ensure that it has the authority to implement these measures when required.

10. The requirements for the portions of the SIP's dealing with episode control are contained in 40 C.F.R. 51.16 (1983), example regulations are presented in 40 C.F.R. 51 Appendix L (1983), and the conditions for EPA approval or disapproval of these portions of the SIP are given in 40 C.F.R. 52.11 (1983).

Essentially these regulations implement the first stage of an episode permit system, while prohibiting the second. While the states allocate the responsibility for the additional control required during episodes, trading priority permits among sources is not provided for. The allocation of responsibility corresponds to the allocation of priority permits. Those forced to cut back implicitly have lower priority permits. Though it would be relatively easy to allow trading among sources, it has not been done. Episode control remains a strict command-and-control approach.

By denying sources the right to trade episode control responsibility, the current regulations force control costs to be higher than they need be though, owing to the absence of studies, the potential cost savings of an episode permit system cannot be calculated. There is no guarantee that the control authority has identified the cheapest quick-response control *sources* much less the cheapest short-response control *measures*. Since sources have nothing to gain by volunteering to use any short-response capability they might have, control authorities are on their own. Only as a result of the compensation that trading provides to cooperating sources could these sources be induced to take on voluntarily the additional responsibility.

Perhaps the most serious flaw in the current regulations is their retardation of the development of new quick-response control technologies. Under the current regulations, sources have an incentive to be as inflexible as possible to avoid being tagged with additional control responsibility. Any apparent flexibility would offer a target of opportunity to control authorities. New quick-response control techniques are viewed as a bother, not a boon, to these sources because they open a door to further control that, in the absence of the new control technologies, would be closed. In the absence of a transferable permit system, adopting these technologies costs money and gains the source nothing in return.

In contrast, episode permit trading allows sources to adopt these quick-response control technologies and to sell their higher priority rights to those who could reduce emissions during episodes only at great cost. The adopting source gets compensated for its willingness to accept the additional responsibility while the compensating source fulfills its legal episode responsibilities more cheaply than would otherwise be possible.

Because episode provisions are rarely invoked, the potential cost savings for this particular further reform may not be as large as for other possible reforms, but they are not negligible. Standby capacity for fuel switching and other quick-response control measures are not cheap, even if rarely used. The associated capital equipment has to be paid for whether used or not and inventories of the substitute fuel have to be maintained. To the extent these are unnecessary expenditures, they could be eliminated by allowing episode permit trades.

SUMMARY

• Air quality measured at specific locations is distributed as a lognormal random variable. Observed readings vary because of underlying variations in meteorological conditions and in emission rates. Ambient standards are usually stated in terms of a long-term average (such as an annual average) or in terms of permissible exceedances of maximum short-term average readings (such as a 3-hour average), or both. The form of the standard is chosen so as to mitigate damage to health (the primary standard) or welfare (the secondary standard).

• Meeting the short-term standards cost effectively means controlling the timing as well as the quantity of emissions. Emissions timing has played no role in the command-and-control regulations. Transferable permit systems can be defined to meet the short-term standards cost effectively. Two different types of permit systems are needed to meet two different types of situations. Periodic permits can be used to control short-term pollution peaks caused by regular, anticipated seasonal or diurnal variations in meteorological conditions. Episode permits can be used to control pollution during those rare, but potentially devastating thermal inversions which can be anticipated only a day or so in advance.

• Though few studies have incorporated the temporal aspects of pollution, the available evidence suggests that significant cost savings may be possible from both periodic and episode control permits. The more stringent the short-term standard, the larger the potential cost savings. These studies have also found that a constant control policy based on a "worst case" condition is frequently not sufficient to avoid violating the ambient standards. Because the true "worst case" depends on emission patterns as well as meteorological conditions, the ambient standards cannot be protected with complete assurance whenever a typical constant-control strategy is adopted.

• As currently written, the Clean Air Act specifically prohibits the use of periodic permits. EPA has not raised the issue of whether the existing command-and-control episode policy might allow trading episode permits. The high costs associated with constant control have tended to discourage the establishment of new short-term standards, opened the door to variances, and delayed attainment.

• Because of the focus on continuous emission control in the Clean Air Act, insufficient attention has been paid to encouraging emission reductions during peak periods and discouraging trades which use emission decreases in off-peak periods to compensate for increases in peak periods. Perhaps the simplest modification to the existing EPA program would be to allow emission reductions occurring in regions designated as nonattainment for violating a short-term ambient standard to be certified as emission reduction credits even when the reductions are not continuous throughout the year.

• Establishing a complete cost-effective periodic control program could increase emission loadings significantly in periods other than peak periods. Should a more limited move toward periodic control be desired, annual average standards could be established to protect against excessive increases in emissions. Procedures would have to be established to prevent emissions from tall stacks escaping this annual average emissions accounting system. With this protection in place, cost-effective temporal control could be pursued without incurring large increases in emission loadings.

• The Clean Air Act prohibition against periodic control eliminates any special incentives for sources to adopt flexible control technologies—those which can reduce emissions relatively cheaply when further control is needed. These malaligned incentives create a bias in the types of control adopted toward those with high fixed and low variable costs, a condition that reduces the ability of state control authorities to secure additional reductions when needed.

• The establishment of a transferable episode permit system would make it easier to identify those sources which can undertake emissions reductions relatively cheaply on short notice. It would also stimulate the development of new relatively inexpensive, short-response control technologies.

REFERENCES

Anderson, Robert J., Jr., Robert O. Reid, and Eugene P. Seskin. 1979. *An Analysis of Alternative Policies for Attaining and Maintaining a Short-Term NO₂ Standard,* a report to the Council on Environmental Quality, prepared by MATHTECH, Inc.

DeBois, Annette Nathanson. 1981. "National Ambient Air Quality Standards Cannot Be Met By Use of Intermittent Controls," *Natural Resources Journal* vol. 21, no. 4 (October) pp. 899–902.

Howe, Charles W., and Dwight R. Lee. 1983. "Priority Pollution Rights: Adapting Pollution Control to a Variable Environment," *Land Economics* vol. 59, no. 2 (May) pp. 141–149.

Larsen, Ralph I. 1971. *A Mathematical Model for Relating Air Quality Measurements to Air Quality Standards,* Environmental Protection Agency Office of

Air Programs AP-89 (Washington, D.C., U.S. Government Printing Office).

O'Neil, William B. (1980). "Pollution Permits and Markets for Water Quality," (unpublished Ph.D. dissertation, the University of Wisconsin-Madison.

———. 1983. "Transferable Discharge Permit Trading Under Varying Stream Conditions: A Simulation of Multiperiod Permit Market Performances on the Fox River, Wisconsin," *Water Resources Research* vol. 19, no. 3 (June) pp. 608–613.

Teller, Azriel. 1970. "Air Pollution Abatement: Economic Rationality and Reality," in Roger Revelle and Hans Landsberg, eds. *America's Changing Environment* (Boston, Beacon Press).

U.S. Dept. of Health, Education, and Welfare. 1970. *Air Quality Criteria for Photochemical Oxidants* (Washington, D.C., U.S. Government Printing Office).

Yaron, Dan. 1979. "A Model for the Analysis of Seasonal Aspects of Water Quality Control," *Journal of Environmental Economics and Management* vol. 6, no. 2 (June) pp. 140–151.

8 / Enforcement

Even the most carefully designed regulatory programs can flounder if the enforcement effort is deficient. Establishing transferable permit systems which allocate the control responsibility cost effectively is of little value if sources regularly fail to comply with the terms of their permits. Ineffective enforcement could undermine the quest for better air quality at lower cost.

The effectiveness of any enforcement program is not only a function of such readily identifiable factors as the size, motivation, and competence of the enforcement staff; the nature of the program makes a difference. Some programs (such as those involving more easily detected violations) are inherently easier to enforce.

Stationary source air pollution control is not one of the easier programs to enforce. Many of the pollutants are invisible to the naked eye and can be measured only with fairly expensive instrumentation. Although the public at large is victimized by pollution, they are so unaware of the dangers that they cannot be relied upon to assist the regulatory authorities by pointing out violations.[1] The regulatory authorities are pretty much on their own.

For our purposes, the main question of interest is how the emissions trading program has affected enforcement. Has enforcement been made easier or more difficult by the program? To what extent do the enforcement properties of the emissions trading program reinforce or limit the ability of the program to accomplish its objectives?

---

1. The Clean Air Act provides for citizen suits, but this provision is rarely used. See 42 U.S.C. 7604.

THE NATURE OF THE ENFORCEMENT PROCESS

Enforcement under the Clean Air Act involves four steps: (1) detecting the violation, (2) notifying the source, (3) negotiating a compliance schedule, (4) applying sanctions for noncompliance when appropriate. Though the states are assumed to have primary responsibility, EPA has the authority to step in and bring enforcement actions against polluters failing to comply with their SIP permits.[2] As a matter of practice the EPA normally takes cases that are particularly complex, that the state has failed to resolve, or that the state has avoided for political reasons.[3]

Detecting the Violation

Two distinct types of compliance have to be verified: (1) initial and (2) continuous compliance. The former involves determining that the plant is in compliance when it commences operation of its control equipment, while the latter involves verifying compliance during the continuous normal operation of the plant throughout its useful life. While a single set of tests conducted at a point in time suffices to determine whether the source is initially in compliance, it is inherently more difficult to verify continuous compliance. Since very few sources are continuously monitored, noncompliance can be verified for most sources only by periodic spot checks.

There are several means used to detect violations, including self-certification by sources, on-site inspections, and direct monitoring of pollutant flows. Self-certification involves reports from the source as to whether or not it is in compliance. These are based on emission levels calculated from process control or fuel data. Continuous compliance is ascertained by annual updates to the initial report submitted when the source applied for its operating permit. Some 36 percent of the major sources judged to be in compliance in 1981 were so judged on the basis of self-certification.[4] Though self-certification is the cheapest means of determining compliance, it may also be the least reliable. Sources have an incentive to place their own compliance behavior in the best possible light.

A somewhat more reliable method involves sending trained control personnel of an inspection visit to the emission site. The inspectors walk

2. The federal and state responsibilities are spelled out in 42 U.S.C. 7413. The reporting requirements which form the basis for enforcement actions are found in 42 U.S.C. 7414.

3. Melnick (1983, p. 197)

4. Crandall and Portney (1983, p. 9). This percentage is down considerably from earlier practice in which self-certification was the dominant source of compliance information.

through the plant checking the operation of the control equipment, conducting smoke density readings, and sampling the fuel as appropriate. Certain types of equipment malfunctions or changes in fuel chemical composition can be easily discovered during these inspections, but others cannot. Normal current practice involves announcing the inspections before they take place, precluding random sampling of the actual operating experience.

The most expensive (and for the majority of sources the most reliable) means of detecting violations involves direct monitoring. This entails the installation of specific emissions monitoring instruments at some key point in the plant, such as a vent or stack. These tests are costly because they involve expensive instrumentation and require specially trained personnel.[5] Since continuous monitoring is a rarity, frequent monitoring visits would be required to obtain a valid picture of the source's continuous compliance status.

Notifying the Source

Notifying the source of a violation initiates the process of regaining compliance. It serves the twin purposes of providing a beginning date for some of the legal procedures and stimulating the source to recognize and to deal with the problem. Upon receiving the notification, the source can unilaterally take steps to reduce emissions sufficiently to return to compliance; it can enter into negotiations with the control authority to define a mutually agreeable response; it can question the validity of the noncompliance finding; or it can simply refuse to comply.

Negotiating Compliance

Unless the source voluntarily submits an acceptable plan showing how it would attain or return to compliance, negotiations are undertaken. These normally attempt to draft a schedule and a means by which the source will attain compliance. Points for negotiation include interim and final deadlines, the types of activities to be undertaken, the kinds of equipment involved, provisions for testing to assure that compliance has been achieved by the deadlines, and the types of penalties to be levied. Most EPA negotiations specify additional controls to be employed in lieu of civil penalties for previous noncompliance.[6]

Litigation

If negotiations do not produce an outcome that is acceptable to the control authority, legal proceedings could be initiated. Because this is an

---

5. Harrington (no date, p. 10) estimates that these tests cost from $2,000 to $10,000 apiece, depending on the complexity and difficulty of the effort.
6. See Melnick (1983, p. 199).

expensive step to take in terms of time and personnel, it is used only when absolutely necessary.

Suits can be filed either by EPA or by the involved state. Suits filed by EPA face a significant bureaucratic hurdle before they even reach the court—the Department of Justice and the local U.S. attorney must be convinced to pursue the case.[7] At the least, this step adds significant delay.

Further delay ensues once the case reaches the court. It takes time for all the pretrial motions to be filed, for the case to be heard, and for a verdict to be rendered. If the court agrees with the control authority, it can issue a regulatory injunction which sets out detailed compliance schedules, and can levy financial penalties for past violations.

THE ECONOMICS OF ENFORCEMENT

This brief description of the enforcement process is probably sufficient to convey one essential point—it could be very misleading to assume perfect enforcement when comparing regulatory approaches. Not only will some violations inevitably go undetected, but not all detected violations result in compliance. At the very least, compliance will be achieved only after a (sometimes substantial) delay.

The degree to which source compliance can be assured affects, and is affected by, the form of the regulatory policy. Policies which may be cost effective with perfect enforceability, but difficult to enforce, may turn out to be less desirable than easier to enforce but less cost-effective policies. To facilitate our understanding of the effect of the emissions trading program on the costs of air quality control, it is useful to consider first how imperfect enforcement would affect the behavior of control authorities and sources.

Cost-Minimizing Source Behavior

A source seeking to minimize costs in an imperfectly enforced regulatory environment must weigh the costs of complying against the costs of not complying.[8] There are two rather distinct categories of compliance costs. The first (and more familiar) category includes the costs of meeting a fixed, predetermined standard, such as expenditures on control equipment, operating costs associated with running the equipment and maintaining it, and the costs of monitoring emissions to verify compliance. The second category includes expenditures to achieve compliance by

7. See Horowitz (1977) for a description of the strained relationship between the Department of Justice and agency attorneys.

8. This section was influenced by the work of Harford (1978), Downing and Watson (1975), and Linder and McBride (forthcoming).

relaxing the standard. These may include lobbying expenditures to amend the regulations or to obtain a variance, as well as litigation expenditures to gain an exemption from, or at least a delay in compliance.

The magnitude of the costs of not complying depend on the likelihood that any violations will be detected and the size of any sanctions levied on identified violators. The lower the likelihood of detection and the outlays required when violations are detected, the more attractive is noncompliance as a means of lowering costs.

By being more precise about how the cost-minimizing source would compare these alternatives, it is possible to create a picture of source decision-making which is quite helpful in tracing out the determinants of noncompliance. The risk-neutral cost-minimizing source faced with imperfect enforcement would choose a level of compliance at which the marginal cost of compliance (including both types of compliance cost) was equal to the expected marginal costs of noncompliance.[9] The expected marginal costs of noncompliance are defined as the likelihood that a violation will be detected and a sanction imposed, multiplied by the marginal outlays required once noncompliance has been established (including fines, legal expenses, etc.)

Perhaps the most important implication of this simple model of regulated source behavior is the overriding importance of the marginal cost of initial compliance and the marginal cost of continuous compliance. As the standards imposed on a source become more stringent, the marginal cost of initial compliance rises, causing expenditures designed to relax the standards to become more attractive. As the marginal cost of continuous compliance rises, the attractiveness of noncompliance as a conscious strategy increases.

With high initial compliance costs, the amount to be saved by avoiding or delaying compliance increases. Delay becomes more attractive with high compliance costs, because the interest earned on the uncommitted funds is larger. These considerations suggest that not only is the political pressure on control authorities to relax standards increased as the stringency of the standards increases, but the expected degree of noncompliance rises as well.[10]

9. Risk-neutral sources by definition are equally satisfied by all events having the same expected payoff. Thus a risk-neutral source would be indifferent between a certain outcome in which it would lose $50 and a lottery in which it had a 50 percent chance of losing nothing and a 50 percent chance of losing $100. A risk-averse firm would perfer the certain outcome, while a risk-loving firm would prefer the lottery. For risk-averse or risk-loving firms, the mix between the detection likelihood and the outlays, once a violation has been established, becomes important, not merely their product.

10. In this model if, as the degree of noncompliance increases, the expected marginal cost of noncompliance rises more slowly than the marginal cost of compliance, it is even possible for actual emissions to increase as the standards are made more stringent. See Harford (1978, p. 33).

Once initial compliance has been demonstrated, subsequent non-compliance need not be as devious as shutting off or by-passing the control equipment, though such actions are not out of the realm of possibility. Noncompliance can also result from a decision to choose a less reliable (and presumably less expensive) control technology or a decision to spend somewhat less on maintenance or upkeep of the equipment than required for continuous compliance.

Downing and Watson (1975) provide an illustrative example of how such a bias in control technology selection could occur with lax enforcement. Two different precipitator technologies could be used to meet the new source performance standards for coal-fired power plants. The first, which they label the flexible technology, uses electronic instrumentation to optimize filtering capacity as discharge electrodes fail over the operating cycle. This optimizing capability provides a means of compensating for any deterioration of performance with use so that the degree of compliance can be maintained. The second, a more inflexible technology, provides no such hedge. Though the inflexible technology is cheaper to run, it is also less reliable over the long run.

Simulating the choices of managers of these plants given less than perfect enforcement, Downing and Watson (1975) find that the source costs are minimized when the inflexible technology is chosen. Though from society's point of view the flexible technology is preferred, from the point of view of the source, the increased reliability of the flexible precipitator is not worth the price as long as enforcement is lax.

There is another different sense in which sources may prefer inflexible technologies. Typically, inflexible technologies have sharply escalating marginal costs as the technology is operated much above its design capacity. Once it is installed, it is very expensive to increase the degree of control with this technology. Though this inflexibility makes the achievement of the standard much more difficult when more control is needed, this inflexibility is an attractive feature to sources. When they plead their case for relaxed standards, the very high cost of further control makes it easier to convince the courts that further reductions are economically infeasible and, therefore, unjustified. With lax enforcement, what is cheapest for the firm is not cheapest for the nation.

Control Authority Responses

The control authority is not powerless in its efforts to assure compliance because it can make noncompliance more expensive. Specifically, it can raise the expected cost of not complying by manipulating the two main elements which make up that cost: (1) the likelihood that violations will be detected and sanctions levied and (2) the level of the sanctions.

In principle, how does the control authority decide how to determine the appropriate level of the expected noncompliance cost? Based on the work of Becker (1968), one method is to set the expected noncompliance cost equal to the expected damage caused by noncompliance. This approach ensures that the source must compensate for, and therefore consider, the harm it causes each time it violates the terms of its permit.

For a given amount of damage caused by noncompliance, this formulation not only requires that higher sanctions be imposed whenever detection probabilities are low, it also requires that payments exceed the damage actually caused. The difference between the value of the damage caused and the size of the monetary sanction is designed to compensate for detection and conviction probabilities that are less than one and to serve as a warning that noncompliance is expensive, even when detection probabilities are low.

This formulation also implies that the expected cost of noncompliance should increase with its duration and intensity. Greater harm triggers larger sanctions. To entertain other designs would undermine the incentive properties of enforcement sanctions. For example, levying the same penalty regardless of the harm caused on all noncomplying sources, say $10,000, would not in general lead to effective enforcement because it would fail to distinguish between slight and gross noncompliance. Once the source has crossed the line into noncompliance, whether intentionally or not, with this penalty structure, it does not reduce its cost by keeping the degree of noncompliance small. In the language of economics, the marginal (or incremental) deterrence for increasing the intensity or duration of noncompliance would be zero.

One of the intentional, but nonetheless controversial, characteristics of this approach is that it provides support for the notion of "optimal noncompliance." By this is meant that for some occasions noncompliance is the expected, even desired, outcome. When the cost of compliance is especially high (such as during an equipment malfunction) and the damage is especially low (such as during periods when the air quality exceeds the standards), the source is expected to violate the standard; this outcome is cheaper, not only for the source, but for society as a whole, because the cost of coming into compliance exceeds the harm caused by noncompliance. Because the source pays the penalty, the malfunction would be corrected, but extraordinary efforts to maintain compliance while the repairs were being made would only be undertaken if they were less expensive than the harm caused.

Even those feeling comfortable with the philosophy behind this basic approach must unfortunately recognize that it has a serious flaw. Reliable estimates of the harm caused by noncompliance are very difficult to acquire. In the absence of these estimates, there is no basis for establishing the level of the sanctions in the Becker-type approach.

An alternative approach having the substantial virtue that it provides a basis for the sanctions to be calculated has been proposed by Drayton (1980). In this approach, any cost savings the source might have achieved by noncompliance are included in the penalty. The objective of this approach is *zero* noncompliance rather than *optimal* noncompliance. By removing any cost advantages the source would accrue from noncompliance, it seeks to eliminate noncompliance altogether.

Though the Drayton proposal fails to recognize the point, assuring zero noncompliance requires that the probability of sanctions being imposed be taken into account to retain the appropriate incentive in the face of incomplete detection.[11] In principle, the noncompliance penalty should be equal to the cost savings from noncompliance divided by the probability that a noncomplying source would be caught and sanctions imposed. This would be equal to the Drayton-type penalty if and only if every noncompliance event was detected and the appropriate penalty imposed in every case. Though this is not beyond the realm of possibility for initial compliance, it is rather hard to accept for continuous compliance. Only by taking the actual probability into account would the expected cost of noncompliance be equal to the cost savings any source could expect to accrue from noncompliance.

In practice, we do not have firm estimates of the probability of detecting noncompliance and levying sanctions. Rather than base the penalty on weakly supported estimates, the noncompliance penalty authorized by the statutes makes no correction for the likelihood of detection.[12] Since this approach could be expected to completely deter noncompliance only if all violations were detected (an uncommon occurrence), this system can be expected to result in more noncompliance than a system involving a correctly defined penalty. It cannot yet be determined with any accuracy how serious a problem this is in practice.

CURRENT ENFORCEMENT PRACTICE

Command-and-Control

How has enforcement worked out in practice? At first glance the effort seems quite successful. The EPA reports that over 90 percent of major polluters (those emitting over 100 tons of a pollutant per year) are in compliance with the terms of their permits.

11. An extended discussion of how this principle might be applied in practice can be found in Levinson (1980).

12. 42 U.S.C. 7420.

That appearance is deceiving, however, because an investigation conducted in 1978 by the General Accounting Office (GAO) of 921 sources considered to be in compliance with their permits revealed 200 (22 percent) of them to be in compliance. In one region, the percentage of supposedly complying sources found to be violating the terms of their permits reached 52 percent.[13]

Problems with the current enforcement process include failures to bring sources identified as not complying into compliance, as well as failures to identify noncomplying sources. The GAO found, for example, in two states visited (New York and Illinois) that of the sources which had been subject to some kind of enforcement action in the period from 1973 to 1977, some 70 percent were still not in compliance by 1977. Still other sources escaped enforcement actions altogether. In Illinois the GAO found that only one-half of the 321 major sources identified as being in noncompliance were faced with enforcement actions by either the states or EPA.[14]

The problem may even be more severe than these data suggest if the air pollution enforcement process shares some of the characteristics of the water pollution enforcement process. A 1984 report by EPA's Inspector General found that it is common practice in water pollution enforcement to respond to violations by issuing administrative orders which effectively sanction the violations by raising the permit limits.[15]

To some extent these discouraging findings flow logically from a reliance on a command-and-control strategy. As the preceding chapters have made clear, a command-and-control strategy raises compliance costs significantly above the least-cost means of achieving air quality goals. High compliance costs provide greater incentives for cost-minimizing firms to seek relaxations of the standards or to accept a larger risk of noncompliance.

Examples of successful standard relaxation are not hard to find. The 1977 Amendments to the Clean Air Act added a provision which said, in effect, that country grain elevators which have a storage capacity of less than 2,500,000 bushels are exempt from any emission standards.[16] Nonferrous smelters were also accorded special treatment.[17]

Even the levels of the primary ambient air quality standards which are supposed to be set purely on the basis of health without considering costs are not immune to this pressure. The smelter industry mounted an attack on a proposed new short-term standard for sulfur dioxide in 1984, argu-

13. U.S. General Accounting Office (1979, p. 9)
14. U.S. General Accounting Office (1981, p. 15).
15. *Inside E.P.A.: Weekly Report*, vol. 5, no. 11 (March 16, 1984), p. 6.
16. 42 U.S.C. 7411(i)
17. 42 U.S.C. 7419

ing that the new standard would deal a "crushing blow" to the industry. The industry argued that EPA is not prohibited by law from considering costs in this case because the Clean Air Scientific Advisory Committee, as well as the EPA staff, have concluded that the health effects do not show a clear advantage of the new standard over the old one.[18] EPA is apparently somewhat sympathetic to this line of argument. Not only is it proposing to consider costs in a limited way in setting a new primary particulate standard, based upon the contention that there is no clear statutory guide as to what constitutes an adequate margin of safety,[19] but EPA is also considering the explicit use of costs in setting secondary standards.[20]

Relaxation can also occur at the state level. In his detailed analysis of enforcement practices in New Mexico, Harrington (1981) found that sources successfully invested in making the regulations applying to them less stringent and in securing variances that permitted delays in installing the control equipment. In one case, by fighting the regulations, Arizona Public Service, the operator of one of the largest emitters in the area, was able to avoid almost all emission reduction requirements for particulates, SO_2 and NO_x.[21] As would be expected, the bulk of the amendments and variances were received by those sources having very high control costs.

Because of its focus on particular technologies of control, the command-and-control approach creates a false sense of security based on two related misconceptions: (1) certifying initial compliance is generally sufficient to guarantee continuous compliance and (2) self-certification of compliance by sources can pick up any exceptions.

Because state enforcement agencies have small staffs, they are forced to adopt short-cut practices. One of these is to allocate a large proportion of manpower to determining initial compliance. When a new or modified source begins operations, relatively extensive source testing is commonly done to verify initial compliance. Subsequent testing is infrequent and less rigorous. Only if initial compliance assures continuous compliance would this approach be an adequate means of enforcing continuous compliance.

Given the degree of noncompliance uncovered in the GAO report, that assumption is of questionable validity. Capital-intensive control technologies can deteriorate with use. The rate of deterioration is an economic as well as a physical variable, depending as it does on the level of maintenance performed on the equipment by the owners. In the

18. *Inside E.P.A.: Weekly Report,* vol. 5, no. 11 (March 16, 1984), p. 2.
19. Ibid., pp.11–12.
20. *Inside E.P.A.: Weekly Report,* vol. 5, no. 8 (February 24, 1984), p. 13.
21. Harrington (1981, pp. 34 and 50).

absence of some kind of continuous surveillance, sources could minimize costs by under-maintaining the control equipment.

The situation is even worse for control strategies where operating costs are particularly high. Since the operating costs can be saved when the control systems are operated at less than full capacity, high operating costs translate into the temptation of high noncompliance.

High operating costs are not uncommon. The Council on Environmental Quality (1980, p. 394) reports that over the period 1979–88 cumulative operating and maintenance costs (excluding mobile sources) for air pollution control were forecast to be $103.9 billion, while cumulative annual capital costs were forecast to be $79.4 billion. Others have found examples of control techniques where operating and maintenance costs are several times annual capital costs.[22]

The presumption that once initial compliance has been achieved continuous compliance would normally follow has not only opened the way for wide-ranging noncompliance, but, perhaps even more important, has diverted attention from the need to develop a more effective system of monitoring continuous compliance. As long as subsequent noncompliance is rare among initially complying sources, expenditures for monitoring continuous compliance would not be needed. Because noncompliance is not rare, more attention has to be paid to monitoring it.

Not all control strategies are subject to these problems. For example, because some strategies recover valuable materials which can be sold for a profit, as long as the value of those materials is sufficiently high (as it sometimes is!), initial compliance offers a high degree of assurance of continuous compliance. In this circumstance the source's costs are minimized by compliance. Unfortunately, a reliance on control technologies yielding valuable by-products is still rare.

The second misconception, viewing self-certification as a sufficient means of assuring continuous compliance, follows from the first. As long as initial compliance implies continuous compliance, there would be little need to check up on sources. Self-certification would produce few temptations and few errors. Conversely, whenever initial compliance does not imply continuous compliance, self-certification becomes a weak means of verifying compliance.

The courts, for their part, have not been particularly helpful. Melnick (1983, pp. 194–195) points out in his thorough review of the courts' role in enforcing the Clean Air Act that though the courts were aggressive supporters of environmental protection when the cases involved defining or expanding the goals of the program, they were very much less supportive of enforcement activities. Because the courts apparently find it more difficult to be supportive when identifiable jobs are at stake, they

22. See Drayton (1980, p. 29) and the sources cited therein.

have been relatively quick to provide relief from regulations on the grounds of economic infeasibility.

The treatment of the smelting industry is illustrative. Two dozen smelters in the western United States account for approximately one-tenth of the nation's sulfur emissions. Recognizing that this industry was economically vulnerable, Congress legislated delayed compliance schedules in 1977. EPA did not publish the smelter regulations required by the act until 1980. Even these rules were overturned by the courts as too strict, leaving the smelters without controls.[23]

Because continuous source testing is expensive, it is unlikely to be employed on all sources. By itself that is not a significant impediment as long as effective sanctions can be imposed whenever noncompliance is detected. With effective sanctions, sources would have an incentive to comply to avoid the penalties on those occasions when violations are discovered.

The penalty system in the Clean Air Act prior to 1977 did not provide much help because the penalties were rarely applied. There seem to be two main reasons: (1) penalties create more friction in the enforcement relationship between sources and the control authorities than almost any other aspect and (2) the procedural burden associated with applying penalties was not small.

Control authorities depend on a certain amount of rapport with the sources they regulate. The absence of rapport can lead to less productive negotiations, to protracted court battles, and ultimately, to fewer enforcement actions because the staff has to spend more time on each case. Penalties are seen by state control authorities as jeopardizing this rapport. Melnick (1983, p. 199), for example, reports that on the basis of his discussions with enforcers, what to do about penalties is often the most difficult issue to resolve during negotiations between sources and control authorities.

There is reason to believe that some of these problems with the use of penalties will be reduced as a result of congressional approval of the noncompliance penalty added by the 1977 amendments.[24] These penalties reduce the procedural burden (and, hence, the reluctance to use them) in two main ways: (1) they provide a clear basis for calculating the penalties and (2) they are civil penalties which can be imposed administratively rather than imposed by the courts.

Little use has yet been made of these penalties. The regulations describing the procedures to be used in calculating the penalties were only

23. See a detailed description of the favorable treatment accorded this industry by the courts in Melnick (1983, pp. 223–227).

24. The statutory basis for noncompliance penalties based on the economic value of noncompliance is 42 U.S.C. 7420.

promulgated in 1980.[25] The EPA was reluctant to impose them with any regularity until their legal basis was established by the courts. EPA data show that only 25 notices of noncompliance were issued under this section of the Clean Air Act from January 1981 to January 1984. Since a favorable decision has now been rendered, more frequent use should be evident in the future.[26]

Even the new noncompliance penalties are not perfect. Not only do they fail to raise expected costs of noncompliance sufficiently to deter noncompliance, as discussed above, but they impose nontrivial procedural burdens.

Sources rarely suffer the imposition of penalties in silence. Though noncompliance penalties can be levied without court involvement, sources may use their rights of appeal to require a full hearing on the record of any petition challenging EPA's issuance of a notice of noncompliance or claiming an exemption. Originally EPA had set up rules which would have denied a hearing in certain "frivolous" cases but the court allowed the petitioners to seek hearings in all cases.[27] Once the battle reaches the courts, even when ultimately settled out of court, time, energy and resources are consumed which may be out of proportion to what is accomplished.

Despite being described by one of its creators as simple,[28] a fifty-six equation system is used to calculate the penalties. This level of complexity invites contestability since there are so many potential points of controversy. One assessed penalty has already been thrown out by the courts on the grounds that the calculation of the penalty was based on spurious and unsupported assumptions.[29] Therefore the verdict on the noncompliance penalty must await further experience with it.

The EPA Emissions Trading Program

How did the establishment of the emissions trading program affect the enforceability of the air pollution control program? This is not an easy question to answer because in some ways it made enforcement easier while in others it made it more difficult.

The most positive effect on enforcement came through its ability to reduce compliance costs. Reduced compliance costs not only mean less incentive for sources to seek relaxation of the standards or to entertain noncompliance as an intentional strategy, they also mean more

25. These regulations can be found in 40 C.F.R. Parts 66 and 67.
26. *Duquesne Light Co.* v. *Environmental Protection Agency* 698 F.2d 456 (1983).
27. 49 FR 9236 at 9237 (March 12, 1984).
28. Drayton (1980, p. 2).
29. *Midland Corporation* v. *Illinois Pollution Control Board,* 14 ELR 20338 (1983).

incentive for the control authorities or the courts to enforce the requirements vigorously.

Because of the high costs of compliance associated with the command-and-control system, it paid sources to invest in relaxing the standards or to delay compliance as long as possible. Several examples of successful use of these tactics, including statutory exemptions, amendments to state regulations, and variances granted either by the control authorities or the courts, have been described in the preceding section. Because investments in delaying and avoiding compliance are themselves expensive, they are only justified when the benefits received are high. When compliance costs are lowered, as they are in the emissions trading program, the benefits from this type of investment (avoided or delayed compliance costs) are diminished. Complying with the regulations becomes relatively more attractive.

The incentives for enforcers to assure compliance are similarly bolstered by lower compliance costs. Control authorities and courts are understandably reluctant to force a source to install very expensive equipment when the cost seems out of line with the accrued benefits. As the simulation models described in chapter 3 make clear, the command-and-control policy distributes the burden in such a way that some firms bear a disproportionately high cost. When these sources bring forth their appeals, they have a stronger case for special treatment than they would in the presence of an emissions trading program. By providing lower cost, alternative ways of attaining compliance (namely, by acquiring emission reduction credits), the emissions trading program makes compliance easier to enforce. The consequences of enforcing the law are less severe than they would be under the command-and-control program.

Not all of the aspects of the emissions trading program facilitate enforcement, however. In particular, by changing the mix of emission points subject to control, this program could make monitoring more difficult, could provide some offsetting increased incentives for non-compliance, and could increase the potential for fraudulent compliance.

THE MONITORING EFFECT. By design, emission trades change the mix of emissions coming from the various points. Because some emission points are inherently easier to monitor than others, changes in this mix can make enforcement either more or less difficult. Some emission trades can and have reduced the difficulties of monitoring. For example, the Du Pont Chambers Works bubble in Deepwater, New Jersey, allows Du Pont to overcontrol 7 large stacks to 99 percent in lieu of 85 percent controls on 119 petrochemical process-fugitive sources. In addition to reducing emissions by some 2,331 tons per year and saving over $12 million in capital costs, this approach made enforcement easier by reducing the number of emission points to be monitored (the process-fugitive sources can now go unmonitored) and by shifting control to those points which are easier to monitor.

Nothing in the current program, however, guarantees that emissions trading would facilitate monitoring. Because the source is not always forced to bear the costs of continuous monitoring, it would not normally give those costs their proper weight in choosing its mix of emission point controls. This creates a bias toward mixes that are more difficult to enforce.

To some extent this monitoring bias is anticipated and countered by the EPA regulations. Trades which offer reductions in fugitive dust for stack particulates are a case in point.[30] Because fugitive dust can come from roadways, construction sites, or other nonpoint sources, particulate reductions achieved by controlling fugitive dust emission points are more difficult to monitor than other trades. Current regulations partially address this problem by requiring a demonstration (either by modeling or monitoring) that emission trades involving fugitive dust are equivalent.[31]

While modeling and monitoring may both be useful for determining equivalency, they are not equally useful for determining continuous compliance. Because modeling generates a one-time estimate accomplished before the fact, it can only address the question of initial compliance. Continuous monitoring provides much more useful information on which to judge continuous compliance. To the extent continuous monitoring is chosen by the source at its expense, current regulations eliminate the bias for this particular type of trade. Because sources will not always make that choice, some bias remains.

THE INCENTIVE FOR NONCOMPLIANCE EFFECT.   Ease of monitoring, however, is not the only enforcement concern triggered by emissions trading. Another is the effect of this program on incentives for noncompliance. By design, the emissions trading program encourages sources to uncover new means of meeting their control responsibilities. Whereas the command-and-control system tended to emphasize capital-intensive control technologies, the emissions trading program opens the way for many noncapital-intensive control technologies, such as process changes, to be instituted. One implication of this shift in emphasis is that operating cost may become an even more important component of total compliance cost under the emissions trading program than it was under the command-and-control approach. Since operating costs can be saved

---

30. For a complete description of the regulatory problems involving the control of fugitive dust, see Probst and Becker (1982).

31. 44 FR 71780 at 71788 (December 11, 1979).

during noncompliance, any rising importance of operating cost could raise the benefits from noncompliance.

The practical implication of this analysis is that the emissions trading program creates an even stronger need for continuous monitoring. Since the current level of continuous monitoring is woefully inadequate for the command-and-control system, it is also inadequate for the emissions trading program. If the emissions trading program serves to stimulate more continuous monitoring efforts, that would turn out to be a substantial boost to the effectiveness of the Clean Air Act.

THE NONEQUIVALENCE EFFECT. The emissions trading program places a new enforcement responsibility on the control authorities—verifying the equivalence of control measures involved in trades. The ability to verify equivalence presumes an accurate estimate of emission decreases at the reduced-control emission point and emission increases at the increased-control emission point, not a trivial task.

The normal starting point for these calculations is the emissions inventory. The emissions inventory *in principle* contains the amount of emissions contributed to the airshed by every individual major source and every category of smaller sources. *In practice* these inventories frequently contain outdated or simply mistaken estimates.

A study conducted by New York City's Division of Air Resources to ascertain the reliability of the emissions inventory illustrates the point.[32] As part of this study, a door-to-door check of heavily industrialized areas was conducted. Some 37 percent of the identified sources that should have been in the inventory were not. Of the sources that had been known previously, some 30 percent had not renewed their certificates to operate as required. Since these renewals are usually used to update the inventory, this is another source of inaccuracy.

Emissions inventories are important because they provide the baseline for calculating emission reduction credits. If the emissions inventories are wrong, the source could qualify for an emission reduction credit even when no actual emission reductions are involved. The problem of defining an appropriate baseline in the absence of an accurate emissions inventory can be illustrated by an actual bubble case.

Monsanto Chemical proposed a bubble for its Chocolate Bayou Facility in Texas in which reduced emissions (200.0 tons per year) resulting from adding an incinerator on one particular vent would be substituted for increased emissions (163.6 tons per year) from eight designated tanks.[33] On the surface, this bubble application would seem to fulfill the

32. Cited in U.S. General Accounting Office (1982, vol. 2, p. 34).
33. 47 FR 55000 at 55501 (December 10, 1982).

equivalency condition, but subsequent probing by the Natural Resources Defense Council has produced some apparent inconsistencies.[34]

While state inventory showed the precontrol level of emissions at the vent to be 105 tons per year, the application assumed 201 tons per year. With an assumed 99.5 percent control of the emissions from this vent, the claimed credit was 200 tons per year whereas the actual reduction, according to the inventory, would be only 104 tons per year. If the inventory is valid, the trade would *increase* actual emissions by some 59 tons per year, not *reduce* emissions by 36.4 tons per year.

Whatever the resolution of this dispute, this case illustrates the importance of having an accurate emissions inventory as a basis for determining equivalency. To the extent actual emissions (as opposed to calculated emissions) increase as a result of an inadequate emissions inventory baseline, air quality deteriorates—a clear violation of the intent of the program. This appears to be one of the most serious threats to the viability of the program—the failure to develop an adequate accounting system to verify the equivalence of emission trades.

SUMMARY

• Enforcement under the Clean Air Act involves four main steps: (1) detecting violations, (2) notifying the source of the violation, (3) negotiating and/or litigating a compliance schedule, and (4) imposing sanctions. Though the states have the primary responsibility for enforcement, EPA has the power to bring enforcement actions against polluters not complying with their SIP responsibilities.

• Cost-minimizing behavior on the part of sources implies weighing the costs of compliance against the costs of noncompliance. The costs of compliance include not only the control costs, but also the costs of seeking relaxed standards through lobbying and litigation. The costs of noncompliance include the resources committed to negotiation or legal defense as well as the payments required by any monetary sanctions imposed.

• The risk-neutral source would minimize its costs when the marginal compliance cost is equal to the expected marginal noncompliance cost. The expected marginal noncompliance cost is the marginal cost of non-

---

34. Letter from David Doniger, Staff Attorney, Natural Resources Defense Council to John Hepola, Chief, State Implementation Plan Section, Environmental Protection Agency (January 10, 1983).

compliance, assuming noncompliance is detected and enforcement actions are taken, multiplied by the probability that the noncompliance would be detected and enforcement actions taken.

• Noncompliance can take several forms, including delays in reaching ultimate compliance, poor maintenance or operation of control equipment, or willful operation of control processes at less-than-necessary capacity.

• Control authorities can attempt to control this behavior by increasing the probability of detection or raising the sanctions imposed on noncomplying sources.

• Two conceptual bases exist for defining appropriate sanctions: (1) optimal and (2) zero noncompliance. The former defines the sanction in terms of the expected harm caused by noncompliance while the latter defines the sanction in terms of the expected cost savings the source would accrue by its failure to comply. Both imply higher sanctions whenever detection probabilities are low.

• Enforcement practices under the command-and-control approach have resulted in a high percentage of noncompliance, running as high as 52 percent in one EPA region according to a Government Accounting Office audit. The high compliance costs associated with the program have also resulted in a number of successful attempts by sources to relax their control responsibilities in the statutes, in state regulations, and in the courts.

• Because of its focus on assuring initial compliance with a set of technology-based emission standards, the command-and-control approach has neglected the need to more effectively monitor and enforce continuous compliance. The evidence is quite clear that initial compliance does not imply continuous compliance.

• Prior to 1977, the noncompliance penalty system was not very effective in encouraging compliance. The penalties were rarely used because control authorities saw them as undermining the rapport between sources and control authorities that is essential if the system is to work and because they imposed a significant procedural burden on the control authorities seeking to use them.

• The 1977 Amendments added a new noncompliance penalty to the enforcement arsenal, based loosely on the zero noncompliance model, which is designed to reduce the procedural burdens. Though this penalty system has flaws, such as not taking detection probabilities into account and relying on a complex calculation formula that invites sources to contest the administrative calculations in the courts, it probably will improve on the current system.

• The emissions trading program can be expected to have mixed effects on enforcement. By lowering compliance costs, it reduces the incentive to seek relaxation of the emission standards and to view non-compliance as a less expensive alternative. On the other hand, by encouraging the development of new sources of control, it opens new possibilities for relying on sources which are more difficult to monitor, for encouraging the control methods with high variable costs, and for allowing trades which degrade the environment.

• Monitoring problems can increase whenever the source does not have responsibility for monitoring, creating a bias toward means of control which are more difficult to monitor. Trades involving reductions in fugitive emissions which do not require monitoring are an example of this problem.

• The tendency for greater compliance resulting from lower compliance cost may be somewhat offset whenever trades substitute control measures with high operating and maintenance costs for those with high capital costs since operating and maintenance costs can be saved by noncompliance. Capital costs are incurred whether the equipment is run at full capacity or not.

• Because the emission inventories maintained by control authorities are currently quite inaccurate, calculating emission reduction credits from the figures in those inventories can lead to trading phantom reductions for actual reductions. Since the practice violates the intent of the program, more effort has to be placed on ensuring that the trades result in equivalency of actual emission reductions.

REFERENCES

Becker, Gary S. 1968. "Crime and Punishment: An Economic Analysis," *Journal of Political Economy,* vol. 76, no. 2 (Mar/Apr.) pp. 169–217.

Council on Environmental Quality. 1980. *Environmental Quality: The Eleventh Annual Report* (Washington, D.C.: U.S. Government Printing Office).

Crandall, Robert W., and Paul R. Portney. 1984. "The Environmental Protection Agency in the Reagan Administration" in Paul R. Portney, ed. *Natural Resources and the Environment* (Washington, D.C.: The Urban Institute Press).

Downing, Paul, and William D. Watson. 1975. "Cost-Effective Enforcement of Environmental Standards," *Journal of the Air Pollution Control Association* vol. 25, no. 7 (July) pp. 705–710.

Drayton, William. 1980. "Economic Law Enforcement," *Harvard Environmental Law Review* vol. 4, no. 1 (Winter) pp. 1–40.

Harford, Jon D. 1978. "Firm Behavior Under Imperfectly Enforceable Pollution Standards and Taxes," *Journal of Environmental Economics and Management* vol. 5, no. 1 (March) pp. 26–43.

Harrington, Winston. 1981. *The Regulatory Approach to Air Quality Management* (Washington, D.C.: Resources for the Future).

————. no date. "Theory vs. Reality in Air Quality Enforcement," unpublished Resources for the Future working paper.

Horowitz, Donald. 1977. *The Jurocracy: Government Lawyers, Agency Programs and Judicial Decisions* (Lexington, Mass., Lexington Books).

Levinson, Michael. 1980. "Deterring Air Polluters through Economically Efficient Sanctions: A Proposal for Amending the Clean Air Act," *Stanford Law Review* vol. 32, no. 4 (April) pp. 807–826.

Linder, Stephen H., and Mark E. McBride. 1984 (forthcoming). "Enforcement Costs and Regulatory Reform: The Agency and Firm Response" *Journal of Environmental Economics and Management*.

Melnick, R. Shep. 1983. *Regulation and the Courts: The Case of the Clean Air Act* (Washington, D.C., Brookings Institution).

Probst, G. L., and R. E. Becker, Jr. 1982. "Escaping the Regulatory Dust Bowl: Fugitive Dust and the Clean Air Act," *Natural Resources Lawyer* vol. 14, no. 3, pp. 541–565.

U.S. General Accounting Office. 1979. *Improvements Needed in Controlling Major Air Pollution Sources* (Report CED-78-165).

U.S. General Accounting Office. 1981. *Clean Air Act: Summary of GAO Reports (October 1977 through January 1981) and Ongoing Reviews* (Report CED-81-84).

U.S. General Accounting Office. 1982. *Cleaning Up the Environment: Progress Achieved But Major Unresolved Issues Remain* (Report CED-82-72).

9 / Evaluation and Proposals for Further Reform

The emissions trading program was in one sense a radical reform of existing policies. Whereas in the command-and-control approach, control authorities had responsibility for both defining the goals and deciding how much emission reduction should be achieved from each individual discharge point, in the emissions trading program the plant managers were allowed great flexibility in choosing the mix of emission reductions used to meet the ambient goals.

The previous chapters have provided a considerable amount of detail on how the program was designed to deal with practical implementation issues and with overcoming resistance from various constituencies.[1] It is now time to coalesce these individual insights into an evaluation of the program as a whole.

EVALUATION

Cost Effectiveness

Measured against the first benchmark, there can be little doubt that the program has improved upon the policy that preceded it. As of December 31, 1983, 61 bubbles had been approved, 6 were proposed, 98 were under review (for proposal) at EPA or the states, and 36 were known to

---

1. For an insider's view of how the program went about the task of constituency building, see Levin (1982).

be under active consideration by companies. Compared with the costs associated with the command-and-control approach, the total estimated capital savings resulting from all approved or proposed bubbles as well as those under development was over $700 million. Over $10 million is being saved every year in reduced operating costs.

These savings were achieved while emission loadings were reduced. Of the 43 bubbles approved or proposed by EPA prior to December 31, 1983 through the state implementation plan (SIP) revision process, fully 27 resulted in substantial air quality improvement.[2] Not only were costs reduced, air quality was improved.

Data on offsets and netting, which are both primarily state-level transactions, are much more difficult to come by. Nonetheless, based on 1983 trend projections from a Department of Energy study completed in 1981, EPA has estimated that roughly 2,500 successful offset transactions have probably taken place. Some several hundred netting transactions were also estimated to have occurred, largely in attainment areas and certain high-growth sections of California.[3]

Cost effectiveness presumes effective enforcement. On this score the evidence is somewhat mixed. Enforcement practices under the command-and-control approach had resulted in high percentages of noncompliance. The high compliance costs associated with command-and-control also resulted in a number of successful attempts by sources to relax their control responsibilities in the statutes, in state regulations, and in the courts. Because of its focus on assuring initial compliance with a set of technology-based emission standards, the command-and-control approach neglected the need to more effectively monitor and enforce continuous compliance. The evidence is quite clear that initial compliance does not imply continuous compliance.

Though it is too early to render a definitive judgment on whether the emissions trading program will ultimately enhance or retard enforcement, a number of influences are now clear. On the positive side, by lowering compliance costs the emissions program has reduced the incentive to seek emission standard relaxations and to view noncompliance as a desirable alternative. Many sources have used the bubble policy as a vehicle for achieving compliance.[4] Due to the possibilities for cost sharing in emissions trading, economic infeasibility should prove less useful as a legal tactic invoked to avoid control responsibilities. On the negative side, by encouraging the development of new sources of control, it has

2. Regulatory Reform Staff (1983, p. 2).
3. Ibid., (pp. 3–4).
4. See the bubbles approved for 3M, 46 FR 41788 (August 18, 1981); U.S. Steel, 48 FR 54347 (December 2, 1983); Fasson-Avery International, 46 FR 61653 (December 18, 1981); and Du Pont, 48 FR 35672 (November 16, 1983).

opened new possibilities for relying on sources which are more difficult to monitor, for making variable control costs (which are more easily saved during noncompliance) a more important part of total control cost, and for allowing trades that deteriorate the environment.

Monitoring problems can increase whenever the source is not responsible for monitoring its emissions, creating a bias toward means of control which are more difficult to monitor. Trades involving reductions in fugitive emissions are an example of a circumstance in which this problem can arise. By imposing monitoring requirements on bubble trades of this type, EPA has sought to minimize the damage.

For a given level of control cost, noncompliance incentives may increase whenever trades substitute high operating and maintenance cost control measures for high capital cost control measures, since operating and maintenance costs can frequently be saved by noncompliance. Capital costs are usually incurred whether the equipment is run at full capacity or not. To date, this problem has not been severe because most of the reductions coming from process modifications involve changes in inputs which are easy to monitor.

Because the emission inventories maintained by control authorities are currently quite inaccurate, calculating emission reduction credits from the figures in those inventories can lead to trading phantom reductions for actual reductions. Since the practice violates the intent of the program, more effort has to be placed on ensuring that the trades result in equivalency of actual emission reductions.

The studies mentioned in chapter 5 found that the poor have borne a disproportionate share of the air pollution control costs resulting from the command-and-control approach, mainly through higher consumer prices. By lowering compliance costs, the emissions trading program has reduced this burden, though the initial burden and the reduction were rather small.

One striking aspect of the emissions trading program is how much more cost effective it has become, compared with what it had been in its earlier years. The program has undergone a considerable evolution in response to statutory and judicial stimuli as well as in response to comments on the regulations by the public at large. In general, this evolution has been liberalizing, removing restrictions which had tended to reduce the cost effectiveness of the program.

EPA had to be cautious to avoid being overruled by the courts. Though based on (and compatible with) general principles in the Clean Air Act, the bubble, netting, and banking programs had no specific statutory authority. In contrast to the offset policy, which was specifically authorized by statute, these three policies were bureaucratic creations. As such, they were vulnerable to strict judicial interpretations.

In view of the need to build a constituency while protecting its flanks from judicial attack, EPA initially proposed heavily circumscribed programs designed to assuage fears and to move slowly. By taking this approach, EPA sought to ensure that the first cases would demonstrate clear, unambiguous benefits and set a useful precedent. At the same time, the number of possible trades would be intentionally limited, giving states time to plan for and become comfortable with the program before any flood of applications overwhelmed them.

The restrictions on the program adopted during this earlier period included a prohibition against emissions banking.[5] The regulations specifically stated, "To allow such 'banking' would be inconsistent with a basic policy of the Act and the ruling—namely, that at a minimum no new source should be allowed to make existing NAAQS violations any worse [41 FR 55526]." At the time that was written, the existence of areas which could not meet the ambient standards by the statutory deadlines was clear, but no procedures had been established for assuring eventual attainment. The concepts of nonattainment areas and reasonable further progress were not yet in the law. The need for such a strict interpretation was diminished by the Clean Air Act Amendments of 1977, which created specific procedures for reaching attainment in nonattainment areas. Since banking could be reasonably integrated into these procedures, it was allowed by the revised regulations promulgated after the amendments were passed.[6]

This was a potentially important boost to the program, since, when banking is not allowed, the incentives for controlling emissions beyond the minimum legal requirements are diminished substantially. Without banking, excess control would be valuable to the creating source only if another source needed an offsetting reduction precisely at the time it was created. One can easily imagine what would happen in more traditional markets such as furniture if the product were confiscated by the state whenever a buyer was not found soon after the product was finished. Less furniture would soon be available. The same principle holds for emission reduction credits. There is absolutely no incentive for sources to undertake additional control voluntarily unless they retain the property right over the emission reduction credit.

Currently all states are encouraged to develop banking programs, though not all states have embraced this opportunity. As of May 1984, EPA had approved only two banking programs (Oregon and Rhode Island) and had proposed to approve two others (Kentucky and Monterey, California). Another eleven states or localities had adopted banking rules, some of which have been forwarded to EPA for review. A

5. 41 FR 55524 at 55529 (December 21, 1976).
6. 44 FR 3274 at 3280 (January 16, 1979).

number of these banks were operational, but the security of the credits is less certain prior to EPA approval.

Despite some attempts by EPA to protect the banked credits, even their rules fall short of fulfilling that objective. While they require the states to designate the owner of the banked credits, the source creating the credits need not always be designated the owner.[7] Confiscation remains a real possibility.

The states for their part have certainly not paid sufficient attention to the need to protect banked credits either. Maine rules, for example, allow the source to retain its property right over banked emission reduction credits for only two years. Following the two-year period, the use of the credits is up to the regulatory authority.[8] Oregon confiscates any credits resulting from permanent source shutdowns or curtailments other than those used by a replacement facility within one year.[9]

This myopic strategy is ultimately self-defeating because it eliminates, or at least reduces, any incentive the source might have had for voluntarily undertaking further reductions. Confiscating banked emission reduction credits may be a cheap source of emission reductions early in the game, but as the treatment of these reductions becomes clear to those who will ultimately create the banked credits, fewer credits are created. When this happens, the control authority will be forced to return to the command-and-control strategy of seeking larger retrofit controls from sources, a strategy which is both more expensive and more difficult to enforce than securing greater reductions at the time of construction, modification, or expansion.

One source of cost ineffectiveness embedded in the early versions of the program which has persisted even in more recent versions is the minimum control thresholds imposed on new or major modified sources. In a cost-effective allocation of control responsibility, these standards would establish a baseline control responsibility, with sources encouraged to fulfill that responsibility in the cheapest way possible, *including the acquisition of either internal or external emission reduction credits.* Compliance would be achieved whenever the combination of reductions achieved by installing control equipment plus emission reduction credits was equal to the baseline control responsibility. The current emissions trading program forbids the use of acquired emission reduction credits as a means of meeting LAER, BACT, or NSPS standards. In general,

---

7. "While the source creating the ERC will generally be its owner, the state could, as part of its rule, reserve ownership of certain classes of ERCs to itself or local governments." 47 FR 15076 at 15084 (April 7, 1982).

8. Maine Bureau of Air Quality Control, *Growth Offset Regulation*, chapter 113, section 4.

9. Oregon Administrative Rules, chapter 340, division 20, section 265.

the only acceptable means of compliance is to secure the designated reduction from each of the specified emission points; no trading is possible. Only RACT standards can routinely be satisfied with emission reduction credits.

Recently EPA has begun to move ever so slightly in the direction of more flexibility in defining these standards. In promulgating its volatile organic compound new source performance standard for the pressure-sensitive tape and label surface coating industry, it applied a production-line standard rather than a standard for each discharge point.[10] In essence, this definition allows the managers of these plants to choose any mix of controls on the production line to meet the stipulated reduction rather than have the mix dictated by EPA. This is the first example of allowing flexibility in how the NSPS is met.

It is important to realize that this approach still falls short of being cost effective by not allowing all possible alternative reduction sources to be considered. Explicitly ruled out are possible reductions from elsewhere in the plant not covered by the NSPS and reductions coming from other plants. Therefore, while it exploits some of the potential cost savings offered by emissions trading, it does not exploit them all.

Prior to this new approach to setting an NSPS, the netting program provided the only current flexibility in applying these standards. This program allows modified plants without significant emission increases to avoid the new source review process entirely. To remain below the significant increase threshold that triggers new source review, sources are allowed to offset potential increases with reductions elsewhere in the plant. Successful use of the netting program does not imply that sources escape all emission standards, however. Those modified sources escaping the new source review process still have to reduce emissions sufficiently to meet the new source performance standard (as opposed to BACT or LAER). Acquired emission reduction credits cannot be used to meet that less stringent standard.

There were two reasons for attempting to increase the flexibility in the new source review process. First, the process itself was very cumbersome and viewed by many as unnecessarily complex.[11] Long permitting delays, a source of tremendous aggravation among plant owners, were common. Second, the cost imposed on new sources was very high, resulting in slower rates of modernization than necessary.

The obvious way to deal with this problem would be to reduce the permitting burden by streamlining the procedural requirements while

10. 48 FR 48368 (October 18, 1983).

11. For a description of the complexities of the permitting process, see National Commission on Air Quality (1981) pp. 163–165.

allowing sources great flexibility in how their control responsibilities could be met. Because the Clean Air Act leaves little room for such flexibility, the netting program was established as a possible, if less desirable, alternative. Significantly, it allowed qualified sources not only to avoid the LAER or BACT requirement, but also to avoid the need to procure offsets in nonattainment areas.

This is a curious means of implementing the act. Instead of making compliance with the emission reductions sought by the act easier, this approach allows them to be circumvented. No doubt some flexibility was needed in light of how cumbersome the permitting process turned out to be, but the Clean Air Act leaves little room for flexibility. Created by a bureaucracy committed to flexibility despite inflexible statutes, the resulting program was perverse; the flexibility provided was achieved by circumventing the new source review program rather than making compliance with it easier.

Other early program restrictions have proved easier to liberalize. The original bubble policy allowed only trades among discharge points within a single plant.[12] Multiplant bubbles were not allowed. By reducing the set of trading opportunities, this policy eliminated many potential sources of significant cost savings. The final policy eliminated this restriction, expanding the potential set of trading opportunities to a large degree.[13]

Liberalization has also characterized the evolution of administrative procedures. Originally the bubble policy could be used only if the approving state included the intended trade in a formal revision to its SIP. Because the SIP approval process is the primary means by which EPA exercises its responsibility for assuring state compliance with the Clean Air Act, SIP revisions are bureaucratically cumbersome. When the Reagan administration took office, for example, a backlog of some 643 proposed changes in SIPs were awaiting approval by EPA.[14] Because any SIP revision has to fulfill a large number of procedural requirements, state control authorities are reluctant to file revisions except when absolutely necessary. Requiring bubble trades to be approved through SIP revisions was a surefire way to limit the amount of interest in the program state control authorities might have had. During 1980 EPA significantly lowered this procedural burden[15] by approving a generic rule for a volatile organic compound (VOC) bubble drawn up by New Jersey. Other states were invited to follow suit. Though initially this invitation included only VOCs, it was subsequently extended to include other criteria pollutants as well.[16] By approving in advance the generic rules

12. 44 FR 3740 at 3741 (January 18, 1979).
13. 44 FR 71780 (December 11, 1979).
14. Crandall and Portney (1984).
15. 45 FR 77459 (November 24, 1980).
16. 47 FR 1507 (April 7, 1982).

states intended to use to govern possible trades, EPA eliminated the need for states to obtain SIP revision approval for each bubble trade. As long as the trades conformed to these rules, no SIP revision was necessary.

This was a major change because it allowed state control authorities to see the bubble policy as something other than a procedural nightmare. As of December 1983, some forty-nine generic bubble applications were under various stages of development or approval as states began to take advantage of this new flexibility. Since with generic rules the policy could be applied with much greater rapidity and much less red tape, by April 1984 the number of bubbles approved over a comparable period under the New Jersey generic rule exceeded those approved directly by EPA for the nation as a whole.[17]

This administrative reform was not implemented without cost. One condition for EPA approval of state generic rules is that emissions not increase as a result of any approved trade.[18] This is more restrictive than the condition applying to SIP revisions. The SIP revision condition is:

> Trades in such areas which increase total emissions can generally occur only as individual SIP revisions in which the state either demonstrates that the trade is consistent with RFP [Reasonable Further Progress] or revises RFP as part of the proposed SIP revision. EPA will approve such revisions as amendments to the SIP, provided they comport with ambient air quality standards and reasonable further progress. [47 FR 15076 at 15082 (April 7, 1982)]

In other words, no trade resulting in increased emissions can be approved under a generic rule even if it were to result in substantially enhanced air quality at those receptors triggering nonattainment and it posed no threat to the attainment status of other receptors. For that class of cost-effective trade, the procedural burden remains high.[19]

The blanket prohibition against emission increases in generic rules was not the only case where cost-effective interplant trades were discouraged. Initially EPA required ambient air quality modeling at the expense of the source for every multiplant bubble trade.[20] This requirement was subsequently relaxed.[21] No modeling is required for VOC or NO_x trades.

17. Telephone conversation with Michael Levin, Director of the Regulatory Reform Staff, EPA, April 3, 1984.

18. 47 FR 15076 at 15078 (April 7, 1982).

19. EPA shows some signs of willingness to reduce this burden. See 48 FR 39580 at 39583, footnote 11 (August 31, 1983).

20. "Where the alternative strategies involve more than one plant, EPA will always require air quality modeling to demonstrate that the increases and decreases in plant emissions will not adversely affect air quality in the area affected by the sources." 44 FR 71780 at 71783 (December 11, 1979).

21. 47 FR 15076 at 15082 (April 7, 1982)

For other pollutants, as long as the relevant sources are located in the same immediate vicinity and no increase in emissions occurs at the source with the lower effective plume height, only the effect of the emissions changes for the two trading sources needs to be modeled. This screening type of modeling is used to ensure that the trade does not cause a significant air quality impact (defined in terms of a concentration increase threshold) at the *receptor* of maximum predicted impact. (Screening modeling cannot identify specific impacts on *monitors* since the model makes no provision for including the effects of all other sources.) Full dispersion modeling, considering all sources in the area of impact, is required by the EPA only if net emissions would increase as a result of the trade or if the trade would have a significant impact on the air quality at the receptor showing maximum ambient impact.

This tailoring of the degree of modeling required to the need for it was a large step forward. Because full-scale modeling is expensive and time-consuming, its use should be reserved for those circumstances requiring it. To do otherwise is to burden multiplant bubbles (or even single plant bubbles if the discharge points are more than 250 meters apart) with an unnecessary requirement which diminishes any interest in distant trades.

Even the existing requirements are excessively severe. Modeling is currently required whenever a trade will significantly affect any receptor *whether or not that increase would trigger a violation of the standard.* Even trades shown by limited modeling as significantly *improving* air quality at the most polluted receptors are subject to the full modeling requirement whenever emissions increase. This restriction discourages trades which could lead to better air quality and earlier compliance in a cost-effective manner.

The problem in overcoming this restriction is that while screening models are inexpensive, they are incapable of predicting ambient standard violations. Full dispersion models, which are capable of predicting violations, are expensive for sources to construct because they must include information on all emissions, not merely those from sources involved in the trade.

Fortunately the problem is soluble because while the fixed costs associated with setting up a full dispersion model are high, the marginal costs of running that model or updating it are relatively low. No source is interested in bearing the initial cost because, for its limited use of the model, the cost is prohibitive. However, since the model could be used to analyze a whole sequence of trades, the cost would be justified when its use for all sources is considered. Thus the problem is not that dispersion models are impractical, but that new means of financing their construction and use need to be worked out.

One particularly serious shortcoming of the emissions trading program involves the definition of the control responsibility baseline. This baseline is important because it serves as the basis for distinguishing those emission reductions which are surplus (and, therefore can be banked or traded) from those that are not. Because the emissions trading program was superimposed on a command-and-control approach, the command-and-control allocation served as the baseline.

The problem arises because historically there was not one baseline, but two—allowable and actual emissions. Traditionally in the command-and-control approach, uniform emission standards were defined for categories of sources (rather than for individual sources), with the standards being set so as to ensure that even the most difficult-to-control source within the category could comply. For the other sources within the category, actual emissions frequently turned out to be considerably lower than allowable emissions. With an allowable emission baseline, the difference between allowable and actual emissions could be certified as an emission reduction credit; with an actual emissions baseline, it could not.

To make matters worse, some states used different baselines for different purposes. In these states, allowable emissions were the base for operating permits, but the lower actual emissions were used to demonstrate progress toward attainment. To permit an allowable emissions baseline when actual emissions were being used to demonstrate attainment could interfere with progress toward attainment, since a trade could result in increased actual emissions even when allowable emissions were held constant.

From a cost-effectiveness point of view, the crucial characteristic of any baseline is consistency. The same baseline should be used to define the operating permits, to construct the emissions inventory used to demonstrate attainment or progress toward attainment, and to define surplus reductions suitable for certification as emission reduction credits. Only with a consistent air quality accounting system can the ambient standards be adequately protected, but the current system is not consistent.

The most serious omission of the emissions trading program lies in its failure to consider adequately the timing of emissions. Meeting the short-term ambient standards cost effectively means controlling the timing as well as the quality of emissions. The high costs associated with using a continuous control strategy to pursue a short-term standard have tended to discourage the establishment of new short-term standards, have opened the door to variances, and have delayed attainment.

The Clean Air Act prohibition against periodic control eliminates any special incentives for sources to adopt flexible control technologies, those which can reduce emissions further as needed at reasonable cost. These malaligned incentives create a bias in the types of control adopted toward those with high fixed and low variable costs, a condition that makes post-installation additional control very difficult.

This analysis suggests that the implementation penalty, the cost savings sacrificed as the price of gaining acceptance, was initially very high, but has been substantially reduced as a result of recent liberalization initiatives. The very large numbers of consummated bubble and offset trades have reduced compliance costs substantially. Most have improved air quality, not merely held it constant.

The flexibility inherent in the program has been undeniably valuable. New sources which were prohibited from entering nonattainment areas under the command-and-control policy can now enter those areas under the terms of the offset policy without jeopardizing progress toward attainment. In addition, stimulated by the cost-savings possibilities, sources have uncovered a large number of ways to rearrange the government-dictated mix of controls that are beneficial both to air quality and to the cost of control.

Yet those features of the program which do not promote cost effectiveness have taken their toll. The number of transactions is significantly smaller than our empirical models have led us to expect. Particularly underrepresented are transactions among different firms. Only two of the thirty-seven SIP revision bubbles approved prior to 1984, for example, involve an interfirm transfer.[22] There is little doubt that these trades have been discouraged by those aspects of the program which increase the administrative burden beyond reasonable bounds.

Another unfulfilled expectation is the degree to which the program has stimulated new innovative ways to control pollution. In a review of those bubbles approved prior to the end of 1983, only four examples of innovative control strategies were uncovered.[23] All of these involved the replacement of solvent-based processes (which are high in volatile organic compounds) with water-based processes. Most of the remaining transactions involve changing the mix of conventional technologies or changing fuels. Given the long lead time associated with the development of new techniques and the rather short history of the program, it is quite possible that this expectation will eventually be fulfilled as the program matures.

Speed of Compliance

Cost effectiveness and enhancing the speed of compliance are compatible objectives. By lowering compliance cost, emissions trading

---

22. These two bubbles involved a transfer of leased emission reduction credits from the International Harvester Company to General Electric (47 FR 1291) and from B.F. Goodrich to the Borden Chemical Company (47 FR 20125).

23. The innovative bubbles include McDonnell Douglas, 46 FR 20172 (April 3, 1981); 3M, 46 FR 41778 (August 18, 1981); Fasson-Avery International, 46 FR 61653 (December 18, 1981) and U.S. Steel, 48 FR 54347 (December 2, 1983).

makes emission reductions easier to secure without placing sources in economic jeopardy. Control authorities are less likely to back down from their responsibility to seek attainment when the impact on jobs is smaller. Not only would sources find that litigation is less attractive, since the savings from delay or an overturned standard would be smaller than when compliance costs are high, but courts are less likely to look favorably on a claim for a reduced burden when the costs are not unreasonable. The expectation of enhanced compliance is certainly reasonable.

Unfortunately the data are not sufficiently reliable to permit a definitive judgment on whether this expectation has been fulfilled. Reports by the General Accounting Office indicate that the available compliance data are generally unreliable.[24] Many sources reported in compliance in fact are not. Add to this unreliability the fact that the vigor with which compliance actions have been enforced has not been constant over the program's history, and the uselessness of the data for supporting any systematic test of changes in the speed of compliance is apparent.

Fortunately some limited evidence is available from those bubbles approved through the SIP revision process. The first piece of evidence concerns two trades which have been responsible for a demonstration of attainment in areas where no such demonstration had previously been possible.[25] In both cases allowing cheaper particulate controls to be substituted for the mandated emission reductions resulted in both lower costs and larger total emission reductions. Since the larger emission reductions were located in areas violating the standards, a demonstration of attainment became feasible once those reductions were taken into account.

In addition to being the vehicle by which some nonattainment areas have been able to demonstrate attainment, the bubble policy has also served in many cases as the means by which individual sources demonstrate compliance with their control responsibilities.[26] In most cases by switching to cheaper means of control, sources were able to achieve the required reductions at lower cost, allowing them to come into compliance.[27]

24. U.S. General Accounting Office (1979, p. i).

25. Armco Steel, 46 FR 19468 (March 30, 1981) and Shenango Coke and Iron Works, 46 FR 62849 (August 18, 1981).

26. 3M, 46 FR 41778 (August 18, 1981); U.S. Steel, 48 FR 54347 (December 2, 1983); Fasson Avery International, 46 FR 61653 (December 18, 1981); Du Pont, 48 FR 35672 (November 16, 1983).

27. In one case compliance was achieved at the expense of an increase in actual emissions. This bubble was approved because it lowered allowable emissions in an air quality control region where the attainment demonstration was based on allowable emissions. See Coors (Colorado), 46 FR 17549 (March 19, 1981).

Although lower compliance cost does not automatically mean more rapid compliance, counterexamples seem to be infrequent and unique. For example, the Uniroyal Plastic Products bubble[28] approval contains a provision extending Uniroyal's deadline for compliance, but the extension was specifically granted to allow Uniroyal to substitute an innovative (and presumably superior) process involving waterborne coatings and inks for more additional control processes involving carbon adsorption or incineration.

Despite this evidence that the emissions program has facilitated compliance compared with the previous policy, there are reasons for being concerned that the current program is not fulfilling its potential. One of the ways in which the current emissions trading program may have caused the speed of compliance to fall below what might have been possible is by creating uncertainty about the rules of the game. New programs always create some uncertainty, but the level of uncertainty associated with the emissions trading program seems particularly high. The rules have changed frequently and dramatically. On the positive side, these changes have generally brought the program closer to being cost effective, but on the negative side they have forced sources to shoot at a moving target.

Sources can make decisions about the types of control to adopt as long as the link between their choice and the outcomes is clear. If the rules keep changing, sources may base their decision on one set of rules, only to find a rather different set in place when enforcement begins. Since changing rules alter the nature of the best strategy, when the degree of uncertainty is high, procrastination is often the source's best strategy. By waiting until the rules are clear, rational responses can be constructed with more confidence. While optimal for the source, procrastination is rarely optimal for society at large.

Changing regulations have not provided the only source of uncertainty inhibiting rapid compliance. Court rulings have provided another. In 1981 the EPA attempted to expand the emissions trading program by applying the netting concept to nonattainment areas. This was a change in direction for the program because it allowed sources undergoing major modifications to procure internal emission reductions to escape the new source review process, even if those sources were located in nonattainment areas. According to the new netting rule, as long as the net increase in emissions was "insignificant," the modification could escape the permit, offset, and emission standard requirements which are normally part of the new source review process. "Insignificant" was defined

---

28. Uniroyal Plastic Products (Ohio). Even the fate of this bubble is not clear. Its initial approval was subsequently withdrawn pending further public comment. See 49 FR 6896 (February 24, 1984).

in terms of a threshold which, for the pollutants of interest, varied from 25 to 40 tons per year.

In *Natural Resources Defense Council, Inc.* v. *Gorsuch,*[29] a lower court ruled that the use of netting in nonattainment areas to avoid new source review was contrary to the purpose of the Clean Air Act and it voided the regulations. The decision was worded broadly enough that it created uncertainty about other previously well-established portions of the trading program, such as allowing bubble trades among existing sources in nonattainment areas.[30] This decision created a void which was filled only when the Supreme Court overruled the Appeals court decision. For the twenty-two months following the Appeal court's decision and prior to the Supreme Court ruling, the bubble process came to an almost complete halt. A number of states stopped writing generic rules and approving bubbles in nonattainment areas, preferring to await the Supreme Court decision. Even though that decision upheld the netting program, the various reversals made implementing the regulations difficult in the interim.

Creating uncertainty or reacting imperfectly to it are not the only ways in which the emissions trading program has retarded the speed of compliance below what it might have been. Restrictions on banking and trading have had a similar effect.

Because current regulations fail to take location into account in defining required emission reductions and disallow many trades which could improve the air quality at the most polluted receptors, they tend to make attainment unnecessarily expensive and more difficult to achieve. By failing to allow for the creation of periodic control and episode control transferable permits, the current regulations not only make it more difficult for sources with installed control equipment to meet the standards, they significantly reduce the incentives for sources to adopt new control technologies that are flexible. Flexible control technologies are particularly important in meeting the short-term standards responsible for many air quality control regions receiving the nonattainment designation.

A key factor in the expectation that the emissions trading program would increase the speed of compliance was the projected role for emission banking. Opportunities for banked credits are created whenever a new plant is constructed or a major change in an existing source takes place. Because the LAER, BACT, and NSPS standards are so stringent, the opportunities for new sources to create additional control are

---

29. 685 F.2d 718 (1982). This decision was overruled in *Chevron U.S.A.* v. *National Resources Defense Council, Inc.,* 52 LW 4845 (June 25, 1984).

30. See the excellent description of the potential legal ramifications of this decision on the program in Reed (1982).

limited. The most important additional reduction credits come from existing sources, such as when old capital equipment is replaced, or process changes are undertaken. These opportunities are desirable sources of surplus reductions because it is frequently much cheaper to combine an investment in additional control with a planned improvement in the plant than it is to retrofit control on a plant that has been operating in the same way for decades.

Banking has not yet lived up to its potential. Though successful banking programs exist, many states have chosen not to develop them. Others have allowed credits to be banked provided they are used or transferred in a short period of time. Still others have failed to eliminate the risk of confiscation. Combined with the restrictions on trades and uncertainties associated with changes in the regulations, the deficiencies of the banking program have caused trading to be less vigorous than expected.

In summary, while the emissions trading policy is certainly more cost effective than the command-and-control policy, it is not fully cost effective. Similarly, while on balance it probably has increased the speed with which the ambient standards are being attained, there are ways in which the speed could be further increased. Fortunately the further reforms that would be necessary to take advantage of these opportunities fall within the spirit of the ongoing liberalization movement.

PROPOSALS FOR FURTHER REFORM

One of the advantages of the approach taken in this book, comparing the current allocation with a cost-effective allocation as well as with the command-and-control allocation, is that it offers a menu of reform possibilities. Each aspect of the program which tends to raise the cost of compliance or tends to retard the speed of compliance becomes a reform candidate. To implement these reforms completely would require actions by Congress, by EPA, and by the states.

Measuring Progress

The overriding objectives of the Clean Air Act are attaining the air quality standards as expeditiously as possible in nonattainment areas and preventing significant deterioration in attainment areas. Attainment could be pursued more rapidly and at lower cost if the yardstick used to measure progress could be changed.

Although the Clean Air Act Amendments of 1977 recognized the existence of nonattainment areas and extended the deadlines for compliance, they also placed special conditions on those areas receiving extensions. One such special condition required the control authorities to ensure that reasonable further progress was made toward attaining the standards by the deadlines. Curiously, reasonable further progress was defined not in terms of reductions in *pollution concentration,* which would ultimately be required for attainment of the standards, but rather in terms of *emission reduction,* which is not uniquely related to concentration reductions. "The term 'reasonable further progress' means annual incremental reductions in emissions of the applicable air pollutant . . . which are sufficient in the judgement of the Administration, to provide for attainment of the national ambient air quality standard by the date required in section 172(a) [42 U.S.C. 7501]." Except for volatile organic compounds, this definition has caused a considerable amount of mischief. For the other criteria pollutants, it is not only perfectly possible for air quality to deteriorate at the most polluted receptors while net emissions are reduced, it is also possible for the air quality at the most polluted receptors to improve without net emission reductions. Since they are neither necessary nor sufficient for attaining the ambient standards, annual emission reductions are a convenient but highly imperfect means of attempting to assure eventual attainment.

Defining the reasonable further progress requirement in terms of emission reductions has left a legacy of questionable practices in the emissions trading program, such as requiring greater-than-equal offsetting emission reductions *regardless of where the reductions would occur and the type of pollutant involved.* Defining reasonable further progress in terms of emissions is appropriate for volatile organic compounds and, to a lesser extent, nitrogen oxides, but for nonuniformly mixed pollutants such as sulfur dioxide and particulates, it is wholly inappropriate. To reach the air quality standards as expeditiously as possible for these pollutants, the location as well as the quantity of the emissions must be taken into account. The 1977 requirement, by focusing exclusively on the latter, has failed to provide incentives to produce better air quality as expeditiously as possible.

Because the reasonable further progress requirement was designed to achieve attainment by 1982, for some pollutants (total suspended particulates and SO_2) the passage of the deadline has eliminated its applicability. For VOCs it remains in force for those areas receiving extensions in the attainment deadlines until 1987.

The diminishing applicability of the reasonable further progress requirement does not solve the problem, even for particulates and SO_2. EPA has published new regulations requiring plans that "provide for attainment as expeditiously as possible," but how this test is to be met is left quite vague.[31] In particular, the regulations do not overrule the traditional interpretation, which is defined in terms of emission reductions.

To correct this problem, the progress requirement should be specifically defined in the act in terms of air quality improvements at those monitors where air quality is worse than the ambient standards rather than in terms of emission reductions. This would not only make this requirement more consistent with the objectives of the act, it would promote more rapid attainment of the ambient standards by lowering the cost of compliance.

Gaining Control over Emission Timing

A second desirable midcourse correction to the Clean Air Act involves the removal of current prohibitions against periodic control, particularly for those nonattainment areas in violation of short-term ambient standards. By removing these prohibitions, one additional means of reaching attainment would be added to the arsenal. This addition would be of particular value to those areas where additional assistance in meeting the ambient standards is needed.

These prohibitions were placed in the act as a response to the fear that periodic controls would replace constant control, resulting in substantially higher emission loadings. However appropriate that response may have been when the act was passed, two considerations suggest that the time for modification has arrived. First, the courts have interpreted the current provision so strictly that they have disallowed periodic control even when the sources would reduce total emissions as well as improve air quality in the worst periods.[32] Currently in many nonattainment areas the level of constant control is so stringent that further levels of continuous reduction are very difficult to achieve. Further controls targeted specifically at those periods when the largest reductions in concentrations are needed could be easier to achieve. Therefore in regions classified as nonattainment due to a violation of a short-term standard, periodic control would be a useful supplement to constant control.

Appropriate safeguards already exist to prevent unwanted increases in emission loadings. The ambient standards for most criteria pollutants are stated in terms of both a long- and a short-term average. The short-term averages protect against exposure to short-duration, high pollution concentrations, while the long-term averages control total emission load-

31. 48 FR 50686 at 50694 (November 2, 1983).

32. See *Big Rivers Electric Corp.* v. *Environmental Protection Agency,* 523 F.2d 16 (1975) and *Kennecott Copper Corporation* v. *Train,* 526 F.2d 1149 (1975).

ings. The existence of a long-term average is a safeguard against increased emission loadings because any such increase would count against the long-term average and could trigger violations.[33]

The ambient standards for a few pollutants are defined only in terms of a short-term average. Presumably the decision to avoid a long-term standard reflects a belief at the time the standard was set that only short-term exposures were important in determining the damage caused. However, if for whatever reason, Congress felt that the absence of a long-term ambient standard afforded too little protection against emission loading increases, it could mandate that every pollutant have an ambient standard based on a long-term average. Those pollutants for which short-duration, high-concentration peaks are found to be damaging could add a short-term standard as well.

Historically the prohibition against periodic control was linked to fears about tall stacks, which are capable of exporting emissions far from the source. Tall stacks are a legitimate concern, but they can be regulated without imposing a blanket prohibition on controlling emission timing. It is time to recognize that while these two aspects of pollution control may once have been historically linked, the link is no longer valid. Most periodic control options do not necessitate the use of tall stacks.

Opening the door to periodic control would not only allow faster attainment at lower cost in precisely those areas where attainment is in doubt and costs are already very high, it would also promote the development and adoption of control technologies which could respond flexibly to the varying needs for control throughout the year. Though the need for some tailoring of the degree of control to local conditions has been recognized in the Clean Air Act in the provisions for episode control, this tailoring has been restricted to those infrequent, random occasions when meteorological conditions are particularly severe. The adoption of a periodic permit approach would extend the applicability of the principle to cover regular, seasonal, or daily peaks in concentrations which currently are triggering violations of short-term ambient standards.

The necessity for periodic control measures would seem to apply with particular force to ozone. A large number of areas are designated nonattainment for ozone, a pollutant with an ambient standard based on a short-term average. Ozone formation has a highly seasonal element, with volatile organic compound reductions in the North having a much higher impact in the summer than in the winter. Because of this larger

---

33. If there is a sense that the current level of the long-term average affords too little protection, the current averages could be lowered.

impact, special efforts to seek summer reductions are warranted even if those reductions cannot be sustained all year long. The constant control bias of the Clean Air Act stands in the way of this sensible strategy.

Reinterpreting LAER, BACT, and NSPS Standards

Perhaps the most important proposed reform involves the role of the minimum control standards in the emissions trading process. Currently emission reduction credits cannot be used to satisfy any portion of the NSPS, LAER, or BACT standards. New or expanding sources must adopt the specific emission reductions at each discharge point specified by these standards whether or not that is the cheapest way to meet the statutory objectives. Because these standards change so infrequently, sources tend to be locked into a control strategy which, even if it made sense when the standard was initially promulgated, may not make sense for sources coming on line years later.

The proposed reform would allow these emission standards to be used to define an overall emission reduction for the plant rather than mandatory reductions to be achieved at each specific discharge point. This plant-wide emission reduction could be achieved by any enforceable means the source desires, including the use of internal or external emission reduction credits. The specific technologies used to define the discharge point standards would no longer have to be adopted. This increase in flexibility would not only lower compliance costs, it would tend to increase the viability of the emission reduction credit market by strengthening the demand for emission reduction credits. Equally important, it would also reduce the disincentive that now exists for replacing old, dirty plants by newer, cleaner ones.

Because the current act does not allow emission reduction credits to be used to meet LAER, NSPS, or BACT standards, the only means open to EPA to provide additional flexibility is to allow modified sources falling below significant increase thresholds to escape the regulations through the netting program. A preferable approach would be to focus on compliance with the standards, not escaping from them. The compliance approach would require all sources to meet the overall emission reduction target imposed by the regulation as well as to procure offsets, but it would allow great flexibility in how the overall emission reduction could be achieved. Not only would the source be free to change the mix of controls on various discharge points, it would also be able to satisfy part of the required overall emission reduction with emission reduction credits purchased from other sources.

Modifying the statute to allow the three types of emission standards to be met more flexibly would offer an excellent opportunity to provide statutory recognition for the bubble policy, something it has not previously enjoyed. The bubble policy could be explicitly incorporated as a device to be used as a means of meeting the plant-wide emission reductions established by NSPS, LAER, and BACT as it currently is for RACT. Interpreting the bubble policy in this manner would serve the twin purpose of easing the sources' compliance burden while lowering the reluctance of control authorities to impose those burdens.

Improving the Air Quality Accounting System

As described in the preceding chapters, inadequacies in the air quality accounting system have been responsible for a considerable amount of mischief in the emissions trading program. The development of an adequate air quality accounting system requires that attention be paid to five highly related aspects of any such system.

PERMITS. All major sources should be required to have operating permits specifying allowable emission rates. In contrast to the current situation, these allowable rates would be tailored to individual sources and would correspond as closely as possible to actual rates. These permits would be used as the basis for determining compliance, for defining surplus emission reductions suitable for certification as emission reduction credits, and for the control authority to establish a reliable emissions inventory.

There is a useful state precedent. As part of its permitting requirements, Oregon assigns to each source subject to regular permitting requirements a plant site emission limit.[34] This limit defines allowable emissions and serves as the basis for developing the SIP, for assuring compliance with the progress requirement, and for defining the baseline for any emissions trading. Banked or traded emission reduction credits are recorded as a reduction in the limit while the purchase of emission reduction credits increases the limit by a like amount. Though it is decidedly easier for a state like Oregon, with its relatively few sources, to implement a program like this, all states should move in this direction. The plant site emission limit is the foundation for a consistent air quality accounting system.

MONITORING. Sources have to be monitored to verify conformance with the permit. Where possible, continuous monitors should be used, but where this is not possible, inspections should be random and especially focused on sources with poor compliance histories.

34. Oregon Administrative Rules, chapter 4, division 20, section 340-20-300.

EMISSIONS INVENTORY. Current emissions inventories in many cases are highly inaccurate. Since accurate dispersion modeling depends on accurate emissions inventories, this is a significant problem. The permitting process should be used to gain accurate information on emission rates, with periodic permit renewals serving to keep this information up to date. This information should be computerized for ease of access.

DISPERSION MODELING. Dispersion models provide the means for relating the emissions inventory and changes in that inventory to air quality changes at the monitors for nonuniformly mixed assimilative pollutants. They would tie any changes in the permits resulting from emissions trading to anticipated air quality effects.

Because of the tremendous importance and initial expense in setting up full-scale dispersion models, particularly in nonattainment areas, Congress should mandate and fund the establishment of operating particulate and SO_2 dispersion models in major metropolitan areas. These models would provide the crucial missing element in the current air quality accounting system, tying together the emissions inventory, operating permits, emissions trading, and progress toward attainment. Once operational, these models could be maintained and run at a relatively low cost. User fees could be used to recover the costs of setting up, maintaining, and operating the models.

The current system, which depends on each individual source to finance the construction of a source-specific full dispersion model, is unrealistic. Individual sources rarely benefit sufficiently from full-scale dispersion modeling to invest in it. Even when the expense of full-scale dispersion modeling would be justified over the long run for all sources, under the current system the models would not be developed.

TERA. The final aspect of an adequate air quality accounting system involves the establishment of an automatic mechanism for handling the situation when air quality turns out to be worse than required by the ambient standards. How the additional reductions are allocated is crucial to the smooth achievement of the standards. From the point of view of cost effectiveness and spreading the financial burden, one attractive procedure is based on transferable emission reduction assessments (TERA).[35] Each polluting source would be assessed a proportion of the required additional reduction based on its RACT responsibility. For the purpose of levying the assessments, banked shutdown credits would be

---

35. The TERA concept was developed in Foster (1978). It is described in National Commission on Air Quality (1981, p. 278) and analyzed in Atkinson and Tietenberg (1984).

treated as the emissions of a (departed) source. Any emission reductions remaining after a source meets its assessment would be available for sale to other sources as one means for them to meet their new requirements. Trading would be encouraged.

Since it is not possible to immediately move to the higher levels of control, interim compliance schedules would have to be developed. More rapid compliance could be encouraged with the use of the non-compliance civil penalty already on the books. Sources complying with this interim schedule would pay no penalties.

There is already a model for this approach which could be emulated by other states. As part of its application for a generic bubble rule, New Jersey imposed RACT standards on a wide variety of sources of volatile organic compounds. Even for those sources where specific controls were not identified, a presumptive RACT requiring reductions of 85 percent or better was established.[36] Sources were allowed to use the bubble policy to rearrange the mix of controls among sources as long as the stipulated plant-wide reductions were achieved. Though this was an individual approach to controlling an individual pollutant, there is no reason why it could not be adopted as a general response to any need for greater control. The establishment of presumptive RACT baselines that are consistent with attainment facilitates trading without jeopardizing air quality.

By instituting a reliable means of relating trades to attainment that is recognized by all parties in advance, both sources and control authorities could act in a less risk-averse way. The current reliance on "snatch and grab" tactics could be reduced as the state control authorities' confidence in their ability to reach attainment in a changing environment was renewed. Once these tactics were banished, sources could afford to place more faith in the sanctity of emission reduction credits, and would be more willing to create new credits. The resulting creation of new credits would reduce the cost of attaining the standards and would reduce the political pressure to protect jobs at the expense of the environment that is faced by state control authorities.

Among other effects, creating this regular process for dealing with change would force sources to consider the likely need for subsequent additional emission reductions at the time of starting, expanding, or modifying the production process. With this explicit evolutionary approach understood by all, future flexibility would be a desirable attribute of any control technology. Flexibility is not a desirable attribute currently, since sources that give any evidence of additional control

36. New Jersey Administrative Code, title 7, chapter 27, subchapter 16 (March 1, 1982), table 4, p. 24.

capability are vulnerable to more stringent control standards. Because they are well aware of this threat, under the current system it pays sources to be inflexible, using this inflexibility as a legal defense against all attempts by the state control authorities to impose more stringent standards on them. It also pays them to avoid trades which might, by virtue of uncovering a greater control capability, expose other similar plants under the same ownership to harsher control. The adoption of specific procedures for dealing realistically with the need for additional control, while allowing for flexible responses on the part of sources, would undermine this brinkmanship strategy and would enhance the speed of compliance.

Protecting Banked Credits

If the emissions banking program is to play its expected role, banked credits have to be a secure asset to the firm. Currently this objective is seen by many state control authorities as conflicting with the need to reach attainment, but that need not be the case. The conflict is not inevitable. Using the TERA procedures for change described above, coupled with plant site emission limits and an air quality accounting system based on full-scale dispersion modeling, the control authority can preserve the sanctity of banked credits while discharging its responsibilities to reach attainment. The sanctity of banked credits conflicts with reaching attainment only when banked credits appear to be the only easy source of reductions. Once credible plans for reaching attainment which integrate preserved credits are developed, the conflict disappears.

It was probably not a coincidence that the two interfirm transfers of emission reduction credits under the SIP revision bubble policy prior to the end of 1983 involved banked credits.[37] Properly protected banked credits facilitate trades both by encouraging the creation of surplus reductions and by coordinating the availability of and the need for these credits over time.

Effective banking programs are clearly possible. For example, as of the middle of 1984, the emissions bank in Jefferson County (Louisville) Kentucky had fifteen depositors of banked credits with running totals of 1,670 tons per year of total suspended particulates, 19,511 tons per year of sulfur dioxide, 1,384 tons per year of volatile organic compounds, 322 tons per year of nitrogen oxides, and 481 tons per year of carbon mon-

37. These bubbles are described in 47 FR 20125 (May 11, 1982) and 47 FR 1291 (January 12, 1982).

oxide.[38] Both of the interfirm trades mentioned in the previous paragraph involved credits made available through this banking system. Other states would do well to emulate this model.

Facilitating Multiplant Trades

Since most states follow the EPA policy guidelines so closely, reforming the statutes and the guidelines would, in practice, reform the state rules. The one major exception involves the treatment of distant multiplant trades in some of the states. These control authorities apply a heavy discount to all distant trades, forcing the acquiring source to purchase much larger emission reductions than needed to offset its increased emissions. As a means of dealing with the hot spot (isolated high concentration) problem, this is an excessively costly approach. Not only does it rule out any trades involving constant or increasing emissions, it makes trades among nonproximate sources uniformly more difficult. A selective approach would differentiate trades improving air quality at monitors experiencing exceedances from those which do not.

Compare the approach taken in California, for example, with one based on the proposed net air quality approach. In the latter, the control authority would calculate the air quality impact of the proposed trade using a full-scale diffusion model. If a 1 to 1 offset ratio would improve air quality at the monitors where violations are occurring (without jeopardizing the ambient standards near the purchasing source), an emissions trade would be allowed. If a 1 to 1 offset ratio would trigger a violation *anywhere,* more units of emission reduction would be required for each unit of increase until complete compliance was assured.

With the net air quality rule, the offset ratio is calculated for the involved sources on the basis of their special circumstances, not a universal rule that must apply to all sources. Not all distant source trades deserve such harsh treatment as they receive in the California approach, though some may.

Consider one circumstance where this could make a large difference. Suppose a source in a heavily polluted area wanted to sell emission reduction credits to a suburban source on a ton-for-ton basis. If consummated, the trade would hold total emissions constant, would lower the cost of compliance, and would improve air quality in the most polluted areas. Yet while the net air quality rule would allow the trade, current rules in most states would rule it out. The California rules are

38. U.S. Environmental Protection Agency, *Emissions Trading Status Report* (May 7, 1984).

extremely crude and costly ways of dealing with a problem which is better handled by requiring a net air quality improvement at the most polluted receptors. The widespread availability of full-scale dispersion modeling would make the more reasonable approach possible.

PROPOSALS FOR FUTURE RESEARCH

Our knowledge of how these programs are and could be working is far from complete. This analysis has uncovered a number of areas in which further research could prove particularly helpful.

As the major stationary sources come under control, more attention has to be paid to the area sources. As the work of Spofford (1984) has made clear, how area sources are controlled has a great deal to do with the cost of reaching the ambient standards. The label "area source" refers to any small residential, governmental, commercial, or industrial source. The approaches discussed in this paper currently have no effect on area sources because they typically fall below the emission thresholds of applicability.

It is impossible to ignore area sources in many locations if the standards are to be attained; collectively they comprise too large a proportion of total emissions. In other areas, even if attainment could be accomplished solely by controlling point sources, that would be an expensive way to meet the standards; it would be cheaper to achieve more of a balance between controlling point and area sources.

Controlling area sources will not be an easy task. Not only are they numerous and diverse, ranging from residences to small commercial establishments, but in many cases the owners or operators have very little information about the available control options. More thought has to be devoted to developing administratively feasible, yet cost-effective ways of imposing the appropriate degree of control responsibility on area sources.

A second research need involves the development of inexpensive, accurate means of continuously monitoring major stationary source emissions. Enforcement efforts would be assisted tremendously by a continuous monitoring capability. Because of its expense, continuous monitoring is rarely required. Cheaper means would allow its wider use. Wider use would, in turn, facilitate the verification of continuous compliance.

A third research need involves documenting and quantifying the impact of the remaining new source bias and the degree to which this impact could be reduced by new-source bubbling. Though its existence is not in doubt, we really know very little about how this effect has affected capital replacement and turnover rates, costs, and air quality, much less the degree to which these effects could be changed by allowing more flexibility in the new source review process. This information would be helpful in assessing the urgency of this portion of the reform package.

The final research need would address our woefully inadequate knowledge of the distinction between the long-run and short-run costs of pollution control. Prior to the construction of any facility, the long-run cost curve is the relevant one, offering a large menu of capital investment control possibilities. After the capital investment choice has been made, however, the short-run cost function becomes relevant; changes in the degree of pollution control at that point are limited by the fixed installed capital.

Traditionally, short-run marginal costs of further control rise more rapidly than long-run marginal costs due to the inability to change the capital base in the short run. For policy purposes it is important to know a great deal more about short-run cost functions than we currently do. Our ability to shed light on how much further flexibility in the emissions trading program would increase the cost savings depends on obtaining this kind of information.

CONCLUDING COMMENTS

The U.S. emissions trading program has carved out an enduring role for itself in U.S. air pollution control policy. Although it substantially improved upon the policy which preceded it, it has fallen short of fulfilling the expectations for it in the theoretical and empirical economics literature. Paradoxically, the very attributes which enabled it to gain a permanent place in U.S. environmental policy are responsible for its failure to be fully cost effective.

The role of EPA's emissions trading program can be directly attributed to the manner in which it was implemented. It was not only compatible with the traditional command-and-control approach, it was implemented to make the transition as smooth as possible. Furthermore, the very high compliance costs associated with the traditional approach created large potential cost savings from adopting this reform. Had the command-and-control policy been more cost effective, it is doubtful that the emissions trading policy could have gained the foothold it has.

Although the ability to overlay this program on an existing policy structure was a key to its political success, it has also diminished the cost effectiveness of the program in several specific ways:

• In response to command-and-control regulations, a large amount of expensive, durable capital control equipment had already been installed prior to the inception of the emissions trading program. Since it was powerless to alter those decisions or even to lower the resulting costs, the emissions trading program reduced costs by less than would have been possible if the program had started with a clean slate.

• A particularly unfortunate side effect of overlaying emissions trading on a preexisting command-and-control allocation arises when some sources comply with the initial regulations rather rapidly and others resist. Because the emissions trading option appears late in the game, sources having already installed control equipment are precluded from using the program to their greatest advantage, while those who were able to fend off early expensive standards can reach compliance at a substantially lower cost. In this way the introduction of an emissions trading program rewards recalcitrant sources, which many potential supporters view as patently unfair.

• In many geographic areas, by ignoring source location in allocating the control responsibility, the command-and-control policy reduced emission loadings more than necessary to meet the air quality standards. Due to the bias in the current program against trades that increase emissions (even if they improve air quality at the most polluted monitors), this initial overcontrol has persisted in the current program, lowering the actual cost savings.

• Because emissions trading played no role in the command-and-control policy, control authorities have been slow in creating a completely congenial environment for trades. Some categories of trades (such as interplant trades) have been made inordinately difficult. This attitude has caused fewer transactions and thinner markets than desirable or possible.

• Seeking to protect existing jobs while making transition as smooth as possible, the EPA implemented a grandfathered version of a marketable emissions permit system that has perpetuated a previously existing bias against new sources. Under this policy, new sources not only face more stringent control responsibilities than existing sources, but they have to acquire sufficient emission reduction credits to offset any emissions remaining after the control is applied. By raising the costs of new plants in relation to existing plants, this bias delays the replacement of high-polluting facilities and has blunted the introduction of those innovative control technologies which are embodied in new sources.

• By using the command-and-control allocations as a point of departure for emission trades, the emissions trading program inherited a significant weakness of that system—an inadequate air quality accounting system. As a practical matter, this deficiency has made states more reluctant to approve trades and has opened the door to transactions which degrade air quality, contrary to the intent of the program.

• The notion that firms should have a property right in surplus emission reductions was not a part of the command-and-control system and has been hard for some control authorities to swallow. In the face of this opposition in principle to one of the main tenets of emissions trading, confiscation of created credits is a distinct possibility, destroying much of the incentive to create additional emission reductions.

These flaws must be kept in perspective. In no way should they overshadow the program's very positive accomplishments. On balance, the current emissions trading program reflects a remarkably astute compromise between the desire to promote cost effectiveness on the one hand and the desire to have an administratively simple, yet politically acceptable program on the other.

In what could serve as a rallying cry for the regulatory reform movement, John Henry Cardinal Newman (1801–1890), the noted English theologian, once said, "Those political institutions are best which subtract as little as possible from a people's natural independence as the price of their protection." The emissions trading program certainly meets that test in principle and to a lesser extent in practice. Although this reform loses its utopian luster upon closer inspection, it has nonetheless made a lasting contribution to environmental policy. It is comforting to note that, on occasion, the realm of the possible and the realm of the desirable overlap.

REFERENCES

Crandall, Robert W., and Paul R. Portney. 1984. "The Environmental Protection Agency in the Reagan Administration," in Paul R. Portney, ed., *Natural Resources and the Environment: The Reagan Approach* (Washington, D.C., Urban Institute Press).

Foster, J. David. 1978. "The Size and Price of Emission Offsets," paper presented at the 71st Air Pollution Control Association annual meeting (June).

Levin, Michael H. 1982. "Getting There: Implementing the 'Bubble' Policy," in Eugene Bardach and Robert A. Kagan, eds., *Social Regulation: Strategies for Reform* (San Francisco, Calif., ICS Press).

National Commission on Air Quality. 1981. *To Breathe Clean Air* (Washington, D.C., U.S. Government Printing Office).

Regulatory Reform Staff. 1983. "Status Report on Emissions Trading Activity" (Washington, D.C., U.S. Environmental Protection Agency, December).

Spofford, Walter O., Jr. 1984. "Efficiency Properties of Alternative Source Control Policies for Meeting Ambient Air Quality Standards: An Empirical Application to the Lower Delaware Valley." Discussion paper D-118 (Washington, D.C., Resources for the Future, November).

U.S. Government Accounting Office. 1979. *Improvements Needed in Controlling Major Air Pollution Sources* (Washington, D.C.).

Index

Acid rain, 27
Advisory Council on Executive
  Organization, 1
Aircraft noise, 108
Air dispersion models, 31, 208
Air pollution
  control studies, 42–43 (table)
  enforcement, 168
  regulatory dilemma, 14–16
  types of, 2–3
Air quality
  accounting system for, 120, 122,
    207–210
  and emissions, 72–73, 88
  and market power, 132–133
  modeling, 195–196
  and reasonable further progress rule,
    89
  and single-receptor dominance, 79
  standards for, 3
Ambient permits, 83–84. *See also*
  Emission trading.
  constraints on, 63–64
  cost savings from, 64
  and emission control, 62–64, 65 (table)
  and emission permits, 24
  financial burden of, 104–105 (table),
    107 (table)
  and geographic interdependency, 27
  and information, 30–31

Ambient permits (*cont'd.*)
  and location problems, 74
  markets for, 27, 78–80
  system difficulties, 60–64
  and transaction complexity, 61–62
Ambient standards, 176
  defining, 151–152
  and emission loadings, 158
  timing of, 205
  and trading activity, 73
Anderson, Robert J., Jr., 157
Area sources, and total emissions, 212
Atkinson, Scott, 72, 85
Auction markets. *See also* Emission
    trading; Zero-revenue auctions.
  for chlorofluorocarbons permits, 108
  competitive/noncompetitive, 129 (table)
  and incentive-compatible auctions, 133
  industry expense under, 109 (figure)
  and initial control responsibility,
    98–100
  and market power, 126–133, 138, 148
  price manipulation in, 145–146

BACT. *See* Best available control
    technology.
Banking, 8–9, 56, 191–192, 201–202. *See
    also* Emission trading; Shutdown
    credits.

Banking (*cont'd.*)
  and credits protection, 210–211
  and location rents, 144
Baseline allocation. *See*
  Command-and-control; Control
  responsibility; Emission trading.
Becker, Gary, 174
Best available control technology
  (BACT), 6
  standards, 206–207
Big Rivers Electric Corporation v.
  Environmental Protection Agency,
  161
Binding receptors, and locational
  incentives, 80
Biological oxygen demand, and emissions
  simulation, 156
Bubble policy, 8, 58, 207. *See also*
  Emission trading.
  and air quality monitoring, 195–196
  air quality test in, 87
  approvals, 52–53
  and banked credits, 210
  capital savings from, 189
  EPA approved, 54–55 (table)
  generic bubble applications, 11, 195
  multiplant bubbles, 194
  requirements of, 86
  and shutdown credits, 116
  transactions by pollutant, 56 (table)
Bureaucracy, and emission trading, 52

California, and multiplant trades, 211
Clean Air Act
  and ambient standards, 62
  and constant control, 160
  and emission permits, 72
  enforcement under, 169, 184
  and episodic control, 163
  and flexibility, 194
  1977 Amendments to, 9, 191, 202–203
  objectives of, 202
  penalty system in, 179–180
  and periodic control, 161–163, 165,
    166, 197
  standards relaxation, 176
  and stationary sources, 2–7
Clean Air Scientific Advisory Committee,
  177
Command-and-control, 15–16
  as baseline allocation, 82–83, 197,
    214–215

Command-and-control (*cont'd.*)
  costs of, 44
  defined, 111
  distribution rule from, 122
  emission loadings, 214
  and emission permits, 66–67, 71 (table)
  emissions reduction from, 65 (table)
  emissions timing in, 165
  and emission trading, 48, 214
  employment impacts of, 95–96
  enforcement under, 175–180, 185
  and equity, 190
  financial burden of, 94–97, 106, 120–121
  industry expense under, 109 (figure)
  and information, 35
  and litigation, 32
  and market power, 136
  new source bias of, 99
  price impacts of, 95–96
  simulation studies of, 39–40
  and source location, 71–72
  standards under, 114
Compliance, 170
  continuous, 178
  and cost-minimizing behavior, 184
  costs of, 171–172, 180–181, 184–185,
    200
  self-certification, 178
Constant control, worst case condition in,
  165
Control authority, and compliance,
  173–175
Control costs. *See also* Financial burden.
  and cartelization, 136
  and locational advantage, 109
  and market power, 131–132
  sensitivity of, 134–135
Control responsibility. *See also*
  Command-and-control; Emission
  trading.
  baseline allocation of, 93, 97–102, 110,
    113–115, 142, 147, 196–197
  command-and-control defined, 111
Control technology
  and compliance strategy, 173
  deterioration of, 177–178
Cost sharing, 122
Council on Environmental Quality, 178
Courts
  emission trading rulings, 200–201
  and enforcement, 178–179
Cumulative emission permit, 29

Delaware River Estuary Region, 49
Domenici, Pete, 2
Dow Chemical, and periodic control, 161–162
Downing, Paul, 173
Drayton, William, 175
DuPage River, 155
DuPont Chambers Works, and bubble policy, 181–182

Emission banking. See Banking.
Emission control
  and air-quality improvement, 88
  cost function for, 130
  fixed costs of, 157
  temporal strategies for, 151–152, 153–160
  and transfer coefficients, 156
Emission permits, 14, 19. See also Emission trading.
  and ambient permits, 24
  and command-and-control, 66–67, 71 (table)
  and compliance speed, 31–34
  cost-effectiveness of, 16–30
  financial burden of, 104–105 (table), 106, 107 (table)
  and information, 30–31
  and location problems, 67–70, 71–72, 74
  and long-range transport, 72
  marketable, 214
  and new sources, 73–74
  for nonuniformly mixed assimilative pollutants, 64–74
  and technological change, 33–34
Emission rates, 150
Emission reduction
  and ambient standards, 119–120
  and permit system, 70–71
  surplus, 215
  and zonal permits, 75
Emissions
  and air quality, 72–73
  constant control of, 155–157
  inventories of, 183, 186, 190, 208
  timing of, 204–206
Emission trading, 2, 2, 103. See also Ambient permits; Auction markets; Banking; Bubble policy; Emission permits; Episode permits; Grandfathering; Market power;

Emission trading (cont'd.)
  Netting; Offset policy; Shutdown credits; Subsidies; Zero-revenue auctions; Zonal permits.
  baseline for, 119
  and command-and-control, 16, 48, 214
  compliance speed, 198–202
  constituency for, 11–12
  constraints on, 39–40
  cost effectiveness of, 35, 51, 121, 188–198
  cost savings from, 41–57
  as cost sharing, 112–113
  distribution rule for, 111
  dynamics of, 49
  and emissions timing, 197–198
  and enforcement, 180–184, 186, 189–190
  EPA program for, 7–11, 38, 57, 86–89, 113–120, 122, 141–145, 147, 160–164, 200, 213
  episode control in, 159–160
  and equity, 96–97, 121, 190
  evaluation of, 188–202
  future research on, 212–213
  hypothetical example of, 25–26
  and market power, 125
  and modeling, 196
  multiplant trades, 211-212
  netting portion of, 117
  in nonattainment areas, 111–112
  and nonequivalence effect, 183–184
  perfect markets assumption of, 50–52
  periodic control in, 153–158
  program benchmarks, 12
  reform proposals, 202–212
  restrictions on, 142–143
  spatial dimension of, 88–89
  success of, 215
  and technological developments, 50
  transactions, 40–41, 52–56
Employment impacts, of command-and-control, 95–96
Enforcement, 169–171
  and cost effectiveness, 189
  current practice, 175–184
  economics of, 171–175
EPA. See U.S. Environmental Protection Agency.
Episode permits, 159–160. See also Emission trading.
  cost savings of, 164
Equity, and pollution control, 94

Financial burden
    distribution of, 107–113, 121
    and market power, 130, 132 (table)
    regional, 102–107
Fox River, 156

General Accounting Office (GAO), on
    compliance, 176, 199
Grandfathering, 110, 113, 121. *See also*
    Emission trading.
    distributional costs with, 111
    and initial control responsibility,
        100–102
    and market power, 133–138, 140, 142
    preference for, 102
    price manipulation in, 146
Growth, and environmental protection,
    34, 36

Hahn, Robert, 101, 134, 136, 137
Harrington, Winston, 177
Harrison, David, 108
Hazardous pollutant, standards, 6
Horizontal equity. *See* Equity.

Information, 35
    and regulation, 15

Kennecott Copper Corporation v. *Train,*
    161
Kuhn-Tucker conditions, 18, 23

LAER. *See* Lowest achievable emission
    rate.
Lagrange multiplier, 18
Litigation, 170–171
Locational incentives, and binding
    receptors, 80
Los Angeles
    air pollution control in, 45
    offset ratio, 87
Lowest achievable emission rate (LAER),
    5
    effect of, 118
    standards, 206–207
Lyon, Randolph, 108

McGartland, Albert Mark, 70–77, 85
Maine, and emissions banking, 192
Maloney, Michael, 136
Market power. *See also* Emission trading.
    and baseline allocation, 134
    and competition, 138–141
    and cost distribution, 129

Market power (*cont'd.*)
    and EPA program, 141–145
    and permit price manipulation, 126–138
    and product competitors, 139
    simulations of, 130–131
    types of, 145
Market structure, for permits, 61-62
Melnick, R., 178–179
Meteorological conditions, 150
Minimum control thresholds, 117–119
    and market power, 143
Minimum treatment standard, effect of,
    118 (figure)
Modeling. *See* Simulation models.
Modified pollution offset, 81
    cost savings from, 85–86
Monitoring, 170, 186, 190, 207
    and ambient permits, 61
    and emission trading, 181, 182
    research need, 212
    and violation detection, 169–170
Monsanto Chemical, and bubble policy,
    183–184
Montgomery, David, 97

Natural Resources Defense Council, 2, 184
    and netting challenge, 11
Natural Resources Defense Council, Inc.
    v. *Gorsuch,* 201
Netting, 8, 56, 193. *See also* Emission
    trading.
    transactions, 189
    and U.S. Supreme Court, 11
New Jersey
    generic bubble rule, 209
    volatile organic compound bubble,
        194–195
Newman, John Henry Cardinal, 215
New Mexico, enforcement practices in,
    177
New source performance standards
    (NSPS), 6–7, 206–207
    flexibility in, 193
New sources
    bias against, 119, 147, 212–213, 214
    and command-and-control, 94–95
    flexibility in, 193–194
    and market power, 140, 143–145
    and minimum control thresholds,
        192–193
    and potential jobs lost, 95–96
Noll, Roger, 101, 136

Nonattainment areas, 5–6
and bubble policy, 199
emissions trading in, 115
netting in, 201
and periodic control, 204
and reasonable further progress, 62–63
Noncompliance
incentives, 182–183, 190
optimal, 174
penalty for, 175, 180, 185
Nondegradation offset, 81, 84–85
and offset ratio, 88
Nonuniformly mixed assimilative
pollutants, 35, 60
cost-effective control of, 22–27
studies of, 44
temporal control of, 154
NO₂, short-term standard for, 158
NSPS. See New source performance
standards.

Oates, Wallace, 85
Offset policy, 7–8, 34, 58, 81, 190. See
also Emission trading.
and offset ratio, 84, 211
problems with, 83
requirements of, 86
and shutdown credits, 116
transactions, 53–56, 189
O'Neil, William, 156–157
Oregon
and emissions banking, 192
plant site emission limit, 120, 207
Ozone
ambient standard for, 152
and bubble transactions, 53
concentrations, 150 (figure)
formation of, 154
and periodic control, 162, 205

Palmer, Adele, 108
Peak-hour pricing, 152–153
Periodic controls
emission loading in, 157–158
prohibitions against, 204–206
Permits, 207
as exhaustible resource, 29
and market power, 136
Piceance Basin, 140, 146
Plant closures, 50. See also Banking.
emission reductions from, 116–117
and market power, 143–145

Plant modification, and LAER
requirement, 117
Pollutant concentrations
and long-range transport, 72
variation in, 149–152
Pollution control
capital investment for, 48–49
economies of scale in, 110, 142
equipment installation, 213–214
innovation in, 198
long-run costs of, 213
marginal cost of, 18, 21, 23
operating and maintenance costs, 178
Pollution sources
cost-minimizing behavior of, 171–173
spatial configurations of, 67–70
Prevention of significant deterioration
(PSD), 6
Price manipulation. See Market power.
PSD. See Prevention of significant
deterioration.

RACT. See Reasonably available control
technology.
Reasonable further progress, 89
defined, 202–204
and plant closures, 116
Reasonably available control technology
(RACT)
and bubble transactions, 53
and emission reduction credits, 193
Regulatory Analysis Review Group, 1
Regulatory reform, 1–2, 38
Reid, Robert O., 157
Revenue auctions. See Auction markets.
Russell, Clifford, 49

San Francisco, offset ratio, 87
Seasonality, 41
and emissions, 19, 23–24
Seskin, Eugene P., 157
Shale oil, 140
Shutdown credits, 122. See also Banking;
Emission trading.
and eminent domain purchase, 145
emission reductions from, 116–117
and market power, 147
Sierra Club, 6
Simulation models, 57
and dynamic effects, 49
measurement errors in, 47–50
of zonal permits system, 76–78

SIP. *See* State implementation plans.
Smelting industry
  and courts, 179
  and standards relaxation, 176–177
Spofford, Walter, 212
Stack heights, 205
  and zonal permits systems, 77–78
State, enforcement agencies, 177
State implementation plants (SIP), 3–6
  baseline in, 114–115
  and bubble policy, 11, 194
  and episode control, 163
Stationary source, statutory performance
  standards, 117 (figure)
Subsidies. *See also* Emission trading.
  incentives from, 100
  and initial control responsibility,
    98–100
  and market power, 126–133

TDP. *See* Emission permits.
Technical change, cost savings from, 33
  (figure)
Teller, Azriel, 160
Tietenberg, Thomas, 85
Total suspended particulates, and bubble
  transactions, 53
Trading rules, 80–86
  comparison of, 82 (figure)
Transactions costs, 51
Transferable discharge permits (TDP).
  See Emission permits.
Transferable emission reduction
  assessments (TERA), 208–210
Transfer coefficients, 74–75
  measurement of, 27

Uncontrolled emission rate, 17
Uniformly mixed accumulative pollutants,
  35
  cost-effective control of, 28–30
Uniformly mixed assimilative pollutants,
  35, 53
  cost-effective control of, 17–22
Uniformly mixed pollutants
  studies of, 44
  temporal control of, 154
Uniroyal Plastic Products, and bubble
  policy, 200
U.S. Environmental Protection Agency
  (EPA), 2, 52–53, 113, 125, 191
  "Economic Dislocation Early Warning
    System," 95

Vertical equity. *See* Equity.

Water pollution
  control studies of, 46
  and enforcement, 176
Watson, William D., 173

Yandle, Bruce, 136
Yaron, Dan, 155–156

Zero-revenue auctions, 110, 121. *See also*
  Auction markets; Emission trading.
  and initial control responsibility,
    100–102
  and market power, 137–139
  preference for, 102
Zonal permits. *See also* Emission trading.
  cost penalty in, 70–77
  and number of zones, 77 (table)
  problems with, 78
  systems of, 74–78

About the Author

Tom Tietenberg is a professor of economics and co-director of the program in public policy at Colby College, Waterville, Maine. During 1983–84 he was a Gilbert White Fellow at Resources for the Future. He has written numerous articles on energy and environmental policy and is the author of the textbook, *Environmental and Natural Resource Economics* (Scott, Foresman, 1984).